Gerhard Reinelt

The Traveling Salesman

Computational Solutions for TSP Applications

Springer-Verlag
Berlin Heidelberg New York
London Paris Tokyo
Hong Kong Barcelona
Budapest

Series Editors

Author

Gerhard Reinelt
Institut für Angewandte Mathematik, Universität Heidelberg
Im Neuenheimer Feld 294, D-69120 Heidelberg, Germany

CR Subject Classification (1991): G.2, I.3.5, G.4, I.2.8, J.1-2

ISBN 3-540-58334-3 Springer-Verlag Berlin Heidelberg New York
ISBN 0-387-58334-3 Springer-Verlag New York Berlin Heidelberg

CIP data applied for

Printed in Germany

Typesetting: Camera-ready by author
SPIN: 10475419 45/3140-543210 - Printed on acid-free paper

Preface

More than fifteen years ago, I was faced with the following problem in an assignment for a class in computer science. A brewery had to deliver beer to five stores, and the task was to write a computer program for determining the shortest route for the truck driver to visit all stores and return to the brewery. All my attemps to find a reasonable algorithm failed, I could not help enumerating all possible routes and then select the best one.

Frustrated at that point, I learnt later that there was no fast algorithm for solving this problem. Moreover, I found that this problem was well known as the traveling salesman problem and that there existed a host of published work on finding solutions. Though no efficient algorithm was developed, there was a tremendous progress in designing fast approximate solutions and even in solving ever larger problem instances to optimality. I started some work on the traveling salesman problem several years ago, first just writing demos for student classes, but then trying to find good and better solutions more effectively. I experienced the fascination of problem solving that, I think, everyone studying the traveling salesman problem will experience. In addition, I found that the problem has relevance in practice and that there is need for fast algorithms.

The present monograph documents my experiments with algorithms for finding good approximate solutions to practical traveling salesman problems. The work presented here profited from discussions and meetings with several people, among them Thomas Christof, Meinrad Funke, Martin Grötschel, Michael Jünger, Manfred Padberg, Giovanni Rinaldi, and Stefan Thienel, not naming dozens of further international researchers.

It is the aim of this text to serve as a guide for practitioners, but also to show that the work on the traveling salesman problem is not at all finished. The TSP will stimulate further efforts and continue to serve as the classical benchmark problem for algorithmic ideas.

Heidelberg, June 1994
Gerhard Reinelt

Preface

More than fifteen years ago, I was faced with the following problem in an assignment for a class in computer science. A brewery had to deliver beer to its stores, and the task was to write a computer program for determining the shortest route for the truck driver to visit all stores and return to the brewery. All my attempts to find a reasonable algorithm failed, I could not help enumerating all possible routes and then select the best one.

Frustrated at that point, I learnt later that there was no fast algorithm for solving this problem. Moreover, I found that this problem was well known as the traveling salesman problem and that there existed a host of published work on finding solutions. Though no efficient algorithm was developed, there was tremendous progress in designing fast approximate solutions and even in solving ever larger problem instances to optimality. I started some work on the traveling salesman problem several years ago, first just writing demos for student classes, but then trying to find good and better solutions more effectively. I experienced the fascination of problem solving that, I think, everyone studying the traveling salesman problem will experience. In addition, I found that the problem has relevance in practice and that there is need for fast algorithms.

The present monograph documents my experiments with algorithms for finding good approximate solutions to practical traveling salesman problems. The results presented here profited from discussions and meetings with several people, among them Thomas Christof, Bernhard Korte, Martin Grötschel, Michael Jünger, Manfred Padberg, Giovanni Rinaldi, and Stefan Thienel, not mentioning dozens of further international researchers.

It is the aim of this text to serve as a guide for practitioners, but also to show that the work on the traveling salesman problem is not at all finished. The TSP will stimulate further efforts and continue to serve as the classical benchmark problem for algorithmic ideas.

Heidelberg, June [illegible]
Gerhard Reinelt

Contents

Chapter 1

Introduction

The most prominent member of the rich set of combinatorial optimization problems is undoubtly the traveling salesman problem (TSP), the task of finding a route through a given set of cities with shortest possible length. It is one of the few mathematical problems that frequently appear in the popular scientific press (CIPRA (1993)) or even in newspapers (KOLATA (1991)). It has a long history, dating back to the 19th century (HOFFMAN & WOLFE (1985)).

The study of this problem has attracted many researchers from different fields, e.g., Mathematics, Operations Research, Physics, Biology, or Artificial Intelligence, and there is a vast amount of literature on it. This is due to the fact that, although it is easily formulated, it exhibits all aspects of combinatorial optimization and has served and continues to serve as the benchmark problem for new algorithmic ideas like simulated annealing, tabu search, neural networks, simulated tunneling or evolutionary methods (to name only a few of them).

On the other hand, the TSP is interesting not only from a theoretical point of view. Many practical applications can be modeled as a traveling salesman problem or as variants of it. Therefore, there is a tremendous need for algorithms. The number of cities in practical applications ranges from some dozens up to even millions (in VLSI design). Due to this manifold area of applications there also has to be a broad collection of algorithms to treat the various special cases.

In the last two decades an enormous progress has been made with respect to solving traveling salesman problems to optimality which, of course, is the ultimate goal of every researcher. Landmarks in the search for optimal solutions are the solution of a 48-city problem (DANTZIG, FULKERSON & JOHNSON (1954)), a 120-city problem (GRÖTSCHEL (1980)), a 318-city problem (CROWDER & PADBERG (1980)), a 532-city problem (PADBERG & RINALDI (1987)), a 666-city problem (GRÖTSCHEL & HOLLAND (1991)), a 2392-city problem (PADBERG & RINALDI (1991)), a 3038-city problem (APPLEGATE, BIXBY, CHVÀTAL & COOK (1991)), and of a 4461-city problem (APPLEGATE, BIXBY, CHVÀTAL & COOK (1993)). This progress is only partly due to the increasing hardware power of computers. Above all, it was made possible by the development of mathematical theory (in particular polyhedral combinatorics) and of efficient algorithms. But, despite of these achievements, the traveling salesman problem is far from being solved. Many aspects of the problem still need to be considered and questions are still left to be answered satisfactorily.

First, the algorithms that are able to solve the largest (with respect to the number of cities) problems to optimality are not stable in the following sense: solution times vary strongly for different problems with the same number of cities and there is no function depending on the number of cities that only gives a slight idea of the time necessary to solve a particular problem. Already problems with some hundred nodes can be very hard for these algorithms and require hours of CPU time on supercomputers. And, there is a lot of theoretical knowledge that has not yet gone into implementations.

Second, problems arising in practice may have a number of cities that is far beyond the capabilities of any exact algorithm available today. There are very good heuristics yielding solutions which are only a few percent above optimality. However, they still can be improved with respect to running time or quality of the computed solutions.

Third, requirements in the production environment may make many algorithms or heuristics unsuitable. Possible reasons are that not enough real time or CPU time is available to apply certain algorithms, that the problem instances are simply too large, or that not enough real time or man power is at hand to code a method one would like to apply.

These arguments visualize the potential that is still inherent in the traveling salesman problem.

The present monograph is meant to be a contribution to practical traveling salesman problem solving. Main emphasis will be laid on the question of how to find good or acceptable tours for large problem instances in short time. We will discuss variants and extensions of known approaches and discuss some new ideas that have proved to be useful. Furthermore we will indicate some directions of future research. Literature will be reviewed to some extent, but a complete coverage of the knowledge about the TSP is beyond the purpose and available space of this tract. For an introduction we recommend the book LAWLER, LENSTRA, RINNOOY KAN & SHMOYS (1985) and the survey article JÜNGER, REINELT & RINALDI (1994).

Nevertheless, even without consulting further references the present text is a meant to be a guide for readers who are concerned with applications of the TSP and aims at providing sufficient information for their successful treatment.

We give a short survey of the topics that will be addressed. Chapter 2 covers basic concepts that we need throughout this monograph. This chapter contains an introduction to complexity theory and describes some fundamental data structures and algorithms. The many possible applications of the TSP are indicated in Chapter 3. Of particular importance are Euclidean instances. To exploit the underlying geometric structure we use Voronoi diagrams and convex hulls which are discussed in Chapter 4. A basic ingredient of fast heuristics will be a limitation of the scope of search for good tours. This is accomplished by candidate sets which restrict algorithms to certain subsets of promising connections. The construction of reasonable candidate sets is the topic of Chapter 5. Construction heuristics to find starting tours are given in Chapter 6. Emphasis is laid on improving standard approaches and making them applicable for larger problems. Many of these heuristics are also useful in later chapters. Chapter 7 is concerned with the improvement of given tours. It is shown how data structures can be successfully employed to come up with very efficient implementations. An important issue is covered in Chapter 8: the treatment of very large problem instances in short time. Several types of approaches are presented. A short survey of recent heuristic methods is contained in Chapter 9. Lower bounds are the topic of Chapter 10. Besides variants of known approaches we comment on heuristics for computing Lagrange multipliers. The algorithms described in this text have been successfully applied in an industry project.

We discuss this project in depth in Chapter 11. Chapter 12 addresses the question of computing optimal solutions as well as solutions with quality guarantee and discusses some lines of future research. In particular, a proposal for a hardware and software setup for the effective treatment of traveling salesman problems in practice is presented. The appendix gives information of how getting access to TSPLIB, a publicly available collection of TSP instances, and lists the current status of these problem instances.

In this monograph we will not describe the approaches down to the implementation level. But, we will give enough information to facilitate implementations and point out possible problems. Algorithms are presented on a level that is sufficient for their understanding and for guiding practical realizations.

Extensive room is spent for computational experiments. Implementations were done carefully, however, due to limited time for coding the software, not always the absolutely fastest algorithm could be used. The main point is the discussion of various algorithmic ideas and their comparison using reasonable implementations. We have not restricted ourselves to only tell "success stories", but we rather point out that sometimes even elaborate approaches fail in practice.

Summarizing, it is the aim of this monograph to give a comprehensive survey on heuristic approaches to traveling salesman problem solving and to motivate the development and implementation of further and possibly better algorithms.

Chapter 2

Basic Concepts

The purpose of this chapter is to survey some basic knowledge from computer science and mathematics that we need in this monograph. It is intended to provide the reader with some fundamental concepts and results. For a more detailed representation of the various subjects we shall refer to appropriate textbooks.

2.1 Graph Theory

Many combinatorial optimization problems can be formulated as problems in graphs. We will therefore review some basic definitions from graph theory.
An **undirected graph** (or **graph**) $G = (V, E)$ consists of a finite set of **nodes** V and a finite set of **edges** E. Each edge e has two **endnodes** u, v and is denoted by $e = uv$ or $e = \{u, v\}$. We call such a graph undirected because we do not distinguish between the edges uv and vu. However, we will sometimes speak about **head** and **tail** of an edge. If $e = uv$ then e is **incident** to v and to u. The set of edges incident to a node v is denoted by $\delta(v)$. The number $|\delta(v)|$ is the **degree** of v.
A graph $G' = (V', E')$ is called a **subgraph** of $G = (V, E)$ if $V' \subseteq V$ and $E' \subseteq E$. For an edge set $\overline{E} \subseteq E$ we define $V(\overline{E}) := \{u, v \in V | uv \in \overline{E}\}$. Conversely, for a node set $\overline{V} \subseteq V$ we define $E(\overline{V}) := \{uv \in E \mid u \in \overline{V} \text{ and } v \in \overline{V}\}$. We call the subgraph $G' = (V(\overline{E}), \overline{E})$ **edge induced (by $\overline{E}$)** and the subgraph $G'' = (\overline{V}, E(\overline{V}))$ **node induced (by $\overline{V}$)**.
A graph $G = (V, E)$ is said to be **complete** if for all $u, v \in V$ it contains edge uv. We denote the complete graph on n nodes by $K_n = (V_n, E_n)$ and assume unless otherwise stated that $V_n = \{1, 2, \ldots, n\}$.
Two graphs $G' = (V', E')$ and $G'' = (V'', E'')$ are **isomorphic** if there exists a bijective mapping $f : V' \to V''$ such that $uv \in E'$ if and only if $f(u)f(v) \in E''$, e.g., the complete graph K_n is unique up to isomorphism.
A graph $G = (V, E)$ is called **bipartite** if its node set V can be partitioned into two nonempty disjoint sets V_1, V_2 with $V_1 \cup V_2 = V$ such that no two nodes in V_1 and no two nodes in V_2 are connected by an edge. If $|V_1| = m, |V_2| = n$ and $E = \{ij \mid i \in V_1, j \in V_2\}$ then we call G the **complete bipartite graph** $K_{m,n}$.
An edge set $P = \{v_1v_2, v_2v_3, \ldots, v_{k-1}v_k\}$ is called a **walk** or more precisely a $[v_1, v_k]$–walk. If $v_i \neq v_j$ for all $i \neq j$ then P is called **path** or $[v_1, v_k]$–path. The **length** of a walk or path is the number of its edges and is denoted by $|P|$. If in a walk $v_1 = v_k$ we speak of a **closed walk**.
A set of edges $C = \{v_1v_2, v_2v_3, \ldots, v_{k-1}v_k, v_kv_1\}$ with $v_i \neq v_j$ for $i \neq j$ is called a **cycle** (or k-**cycle**). An edge $v_iv_j, 1 \leq i \neq j \leq k$, not in C is called **chord** of C. The length of

a cycle C is denoted by $|C|$. For convenience we shall sometimes abbreviate the cycle $\{v_1v_2, v_2v_3, \ldots, v_kv_1\}$ by $(v_1, v_2, \ldots v_k)$ and also say that a graph G is a cycle if its edge set forms a cycle. A graph or edge set is called **acyclic** if it contains no cycle. An acyclic graph is also called **forest**.

A graph $G = (V, E)$ is said to be **connected** if it contains for every pair of nodes a path connecting them; otherwise G is called **disconnected**. A **spanning tree** is a connected forest containing all nodes of the graph.

A nonempty edge set $F \subseteq E$ is said to be a **cut** of the graph $G = (V, E)$ if V can be partitioned into two nonempty disjoint subsets V_1, V_2 with $V_1 \cup V_2 = V$ such that the following holds: $F = \{uv \in E \mid u \in V_1, v \in V_2\}$. Equivalently, F is a cut if there exists a node set $W \subseteq V$ such that $F = \delta(W)$.

Sometimes it is useful to associate a direction with the edges of a graph. A **directed graph** (or **digraph**) $D = (V, A)$ consists of a finite set of **nodes** V and a set of **arcs** $A \subseteq V \times V \setminus \{(v, v) \mid v \in V\}$ (we do not consider loops or multiple arcs). If $e = (u, v)$ is an arc of D with **endnodes** u and v then we call u its **tail** and v its **head**. The arc e is said to be **directed from** u **to** v, **incident from** u and **incident to** v. The number of arcs incident to a node v is called the **indegree** of v and the number of arcs incident from v is called the **outdegree** of v. The **degree** of v is the sum of its indegree and outdegree. For a node v the sets of arcs incident from v, incident to v, and incident from or to v are denoted by $\delta^+(v)$, $\delta^-(v)$, and $\delta(v)$, respectively. Two nodes are **adjacent** if there is an arc connecting them.

Most of the definitions for undirected graphs carry over in a straightforward way to directed graphs. For example, **diwalks**, **dipaths** and **dicycles** are defined analogously to walks, paths, and cycles with the additional requirement that the arcs are directed in the same direction.

A digraph $D = (V, A)$ is said to be **complete** if for all $u, v \in V$ it contains both arcs (u, v) and (v, u). We denote the complete digraph on n nodes by $D_n = (V_n, A_n)$.

For each digraph $D = (V, A)$ we can construct its **underlying graph** $G = (V, E)$ by setting $E = \{uv \mid u \text{ and } v \text{ are adjacent in } D\}$.

A digraph $D = (V, A)$ is called connected (disconnected) if its underlying graph is connected (disconnected). D is called **diconnected** if for each pair u, v of its nodes there are a $[u, v]$– and a $[v, u]$–dipath in D. A node $v \in V$ is called **articulation node** or **cutnode** of a digraph (graph) if the removal of v and all arcs (edges) having v as an endnode disconnects the digraph (graph). A connected digraph (graph) is said to be **2-connected** if it contains no articulation node.

To avoid degenerate situations we assume that, unless otherwise noted, all graphs and digraphs contain at least one edge, respectively arc.

A walk (diwalk) that traverses every edge (arc) of a graph (digraph) exactly once is called **Eulerian trail (Eulerian ditrail)**. If such a walk (diwalk) is closed we speak of a **Eulerian tour**. A graph (digraph) is **Eulerian** if its edge (arc) set can be traversed by a Eulerian tour.

A cycle (dicycle) of length n in a graph (digraph) on n nodes is called **Hamiltonian cycle (Hamiltonian dicycle)** or **Hamiltonian tour**. A path (dipath) of length n is called **Hamiltonian path (Hamiltonian dipath)**. A graph (digraph) containing a Hamiltonian tour is called **Hamiltonian.**

Often we have to deal with graphs where a rational number (edge weight) is associated with each edge. We call a function $c : E \to \mathbf{Q}$ (where $\mathbf{Q}$ denotes the set of rational numbers) a **weight function** defining a weight $c(e)$ (or c_e, or c_{uv}) for every edge $e = uv \in E$. (In the context of practical computations it makes no sense to admit arbitrary real-valued functions since only rational numbers are representable on a computer.) The weight of a set of edges $F \subseteq E$ is defined as

$$c(F) := \sum_{uv \in F} c_{uv}.$$

The weight of a tour is usually called its **length**, a tour of smallest weight is called **shortest tour**. The problem in the focus of this monograph is the so-called (symmetric) traveling salesman problem.

(Symmetric) Traveling Salesman Problem
Given the complete graph K_n with edge weights c_{uv} find a shortest Hamiltonian tour in K_n.

□

A symmetric TSP is said to satisfy the **triangle inequality**, if $c_{uv} \leq c_{uw} + c_{wv}$ for all distinct nodes $u, v, w \in V$. Of particular interest are **metric** traveling salesman problems. These are problems where the nodes correspond to points in some space and where the edge weights are given by evaluating some metric distance between corresponding points. For example, a **Euclidean TSP** is defined by a set of points in the plane. The corresponding graph contains a node for every point and edge weights are given by the Euclidean distance of the points associated with the end nodes.
We list some problems on graphs related to the traveling salesman problem which will be referred to at some places.

Asymmetric Traveling Salesman Problem
Given the complete digraph D_n with arc weights c_{uv} find a shortest Hamiltonian tour in D_n.

□

Chinese Postman Problem
Given a graph $G = (V, E)$ with edge weights c_{uv} for $uv \in E$ find a shortest closed walk in G containing all edges at least once.

□

Hamiltonian Cycle Problem
Given a graph $G = (V, E)$ decide if G contains a Hamiltonian cycle.

□

Eulerian Tour Problem
Given a graph $G = (V, E)$ decide if G is Eulerian.

□

Though quite similar, these problems are very different with respect to their hardness. It is a topic of the next section to give a short introduction into complexity theory and its impact on the traveling salesman problem.

2.2 Complexity Theory

When dealing with combinatorial problems or algorithms one is often interested in comparing problems with respect to their hardness and algorithms with respect to their efficiency. Often it is intuitively clear that some problem is more difficult to solve than another problem and that one algorithm takes longer than another algorithm. The work of Cook (COOK (1971)) laid the foundation for putting these questions into an exact mathematical framework. Based on the notion of deterministic and nondeterministic Turing machines it makes the classification of problems as "hard" or "easy" possible and allows the measurement of efficiency of algorithms. Although being only a theoretical model this concept had a great impact on the design and the analysis of algorithms.
For our purposes it is sufficient to introduce the concepts of complexity theory in a more informal manner. If we omit certain subtleties we can take a real-world computer as our computational model and think of an algorithm as a procedure written in some high-level programming language. For a thorough study of complexity issues we recommend GAREY & JOHNSON (1979).
For reasons of exactness we distinguish in this section between "problems" and "instances of problems". A **problem** or **problem class** is a question defined on several formal parameters, e.g., determine whether a graph contains a Hamiltonian cycle or compute a shortest Hamiltonian tour in a weighted graph. If we associate concrete values to these formal parameters we create a particular **instance** of the problem. A particular graph defines an instance of the Hamiltonian cycle problem and a particular weight function $c : E_n \to \mathbf{Q}$ gives an instance of the traveling salesman problem for the complete graph $K_n = (V_n, E_n)$.
These two examples show in addition that we have to distinguish between two types of problems. One type is the so-called **decision problem** which requires a "yes" or "no" answer and the other type is the **optimization problem** demanding to exhibit a solution which optimizes some objective function. We first give the basic terminology which has been defined for decision problems and then show how optimization problems can also be handled within this concept.
The performance of an algorithm has to be measured in some way depending on the "sizes" of the problem instances to be solved. Therefore we associate with each instance I of a certain problem class S a **size** or **encoding length** $l(I)$ which is defined as the number of bits required to represent the actual parameters in the usual binary encoding scheme. If A is an algorithm for the solution of problem S then we define its running time (for instance I) as the number of elementary operations (addition, multiplication, etc.) which have to be executed on a computer to solve instance I. The **time complexity** or **running time** of an algorithm A for a problem class S is then defined as a function $t_A : \mathbf{N} \to \mathbf{N}$ giving for each natural number n the number $t_A(n)$ of elementary operations that the algorithm has to execute at most to solve an instance of size n.
Note, that we assume that arithmetic operations are executed in constant time, i.e., independent of the size of the numbers involved. This is not correct in general, but is feasible in our context.
Of course, in the usual case we will not be able to derive an explicit formula for evaluating $t_A(n)$. On the other hand, we are not interested in concrete values of t_A but rather in

growth of t_A with increasing n. If it is not possible to give the exact rate ve are interested in lower and upper bounds for this rate. In the case of a pro... e are also interested in bounds for the running time of algorithms that are able to solve the problem. We introduce some notations to express knowledge about the rate of growth or bounds on this rate.

Definition 2.1 *Let $f : \mathbf{N} \to \mathbf{N}$ and $g : \mathbf{N} \to \mathbf{N}$ be given.*

(i) We say that f is $O(g)$ if there exist positive constants c and n_0 such that $0 \le f(n) \le c \cdot g(n)$ for all $n \ge n_0$.

(ii) We say that f is $\Omega(g)$ if there exist positive constants c and n_0 such that $0 \le c \cdot g(n) \le f(n)$ for all $n \ge n_0$.

(iii) We say that f is $\Theta(g)$ if there exist positive constants c_1, c_2, and n_0 such that $0 \le c_1 \cdot g(n) \le f(n) \le c_2 \cdot g(n)$ for all $n \ge n_0$. □

The three notations define asymptotic upper, lower, and tight bounds, respectively, on the rate of growth of f. An alternate definition of an asymptotic lower bound is obtained by replacing "for all $n \ge n_0$" in 2.1 (iii) by "for infinitely many n". Asymptotic upper bounds are of practical interest since they give a **worst case running time** of an algorithm. It is usually harder to derive nontrivial asymptotic lower bounds, but we will occasionally be able to give such bounds.
We also use the Ω- and Θ-notation for problems. With the first notation we indicate lower bounds on the running time of any algorithm that solves the problem, with the second notation we indicate that an algorithm with best possible time complexity exists to solve the problem.
An algorithm A is said to have **polynomial time complexity** if there exists a polynomial p such that $t_A(n) = O(p(n))$. All other algorithms are said to be of **exponential time complexity** (although there are superpolynomial functions not being exponential). EDMONDS (1965) was the first to emphasize the difference between polynomial and nonpolynomial algorithms. It is now commonly accepted that only algorithms having a polynomial worst case time complexity should be termed **efficient algorithms**.
We denote by $\mathcal{P}$ the class of decision problems which can be solved by polynomial time algorithms.
The Eulerian tour problem can easily be solved in polynomial time. Using a result from graph theory the following algorithm tests whether a connected graph $G = (V, E)$ is Eulerian.

procedure eulerian(G)

(1) For every $v \in V$ compute its degree $|\delta(v)|$.

(2) If all node degrees are even then the graph is Eulerian. If the degree of at least one node is odd then the graph is not Eulerian.

end of eulerian

This algorithm runs in time $\Theta(n+m)$. Surprisingly, also the Chinese postman problem can be solved in polynomial time (EDMONDS & JOHNSON (1973)).

However, many problems (in fact, most of the interesting problems in combinatorial optimization) can up to date not be solved (and probably are not solvable) by polynomial time algorithms. From a theoretical viewpoint they could be solved in the following way. If the answer to an instance I (of a decision problem) is "yes", then in a first step some string s whose length is polynomial in the input size is guessed nondeterministically. In a second step it is verified that s proves that the problem has a "yes" answer. The verification step is performed (deterministically) in time polynomial both in the length of s and in the size of I. If the answer to I is "no" then there exists no such string and the algorithm is assumed to run forever. E.g., in the Hamiltonian cycle problem the string s could be the encoding of a Hamiltonian cycle (if the graph contains such a cycle); the length of s is polynomial in the input length, and it can be easily verified whether s is indeed the encoding of a Hamiltonian cycle.

Obviously this procedure cannot be realized in practice. The formal model enabling such computations is the so-called **nondeterministic Turing machine**. For our purposes we can think of the instruction set of an ordinary computer enhanced by the instruction "Investigate the following two branches in parallel". The time complexity of a nondeterministic algorithm is the maximal number of elementary steps that is required to solve a decision problem if it has a "yes" answer.

The class of decision problems that can be solved in polynomial time using such nondeterministic algorithms is called $\mathcal{NP}$.

Note the important point that there is an asymmetry between "yes" and "no" answers here. The question of how to show that a decision problem has a "no" answer is not considered in this concept.

An important subclass of $\mathcal{NP}$ consists of the $\mathcal{NP}$-complete problems. These are the hardest problems in $\mathcal{NP}$ in the sense that if one of them is shown to be in $\mathcal{P}$ then $\mathcal{P}=\mathcal{NP}$. Let A be an algorithm for the solution of problem $\mathcal{B}$. We say that a problem $\mathcal{C}$ is **polynomially reducible** to problem $\mathcal{B}$ if it can be solved in polynomial time by an algorithm that uses A as a subroutine provided that each subroutine call of A only counts as one step. A problem is then called **$\mathcal{NP}$-complete** if every problem in $\mathcal{NP}$ is polynomially reducible to it.

The Hamiltonian cycle problem is one member of the broad collection of $\mathcal{NP}$-complete problems (for a derivation of this result see JOHNSON & PAPADIMITRIOU (1985)).

The question $\mathcal{P}=\mathcal{NP}$? is one of the most famous unsolved questions in complexity theory. Since this question has now been attacked for two decades and since $\mathcal{NP}$-complete problems proved to be substantially hard in practice it is commonly accepted that $\mathcal{P}\neq\mathcal{NP}$ should be the probable answer to this question (if it can be decided at all).

We want to emphasize again that our representation is kept on an informal level, and it is intended to give just an idea of the concepts of complexity theory. Especially we have not considered space complexity which measures the amount of storage an algorithm requires.

We now discuss how optimization problems like the traveling salesman problem can be dealt with. With the TSP we associate the following decision problem which can be analyzed using the above concepts.

Traveling Salesman Decision Problem
Given the complete graph K_n with edge weights c_{uv} and a number b decide if there exists a Hamiltonian tour in K_n with length at most b. □

This decision problem is $\mathcal{NP}$-complete (JOHNSON & PAPADIMITRIOU (1985)).
If the traveling salesman problem is in $\mathcal{P}$ then obviously also the corresponding decision problem is in $\mathcal{P}$. An optimization problem having the property that the existence of a polynomial time algorithm for the solution of an associated decision problem implies the polynomial solvability of an $\mathcal{NP}$-complete problem, is said to be **$\mathcal{NP}$-hard.**
On the other hand, assume there exists a polynomial time algorithm for the solution of the TSP decision problem. If all edge weights are integral and the largest weight in absolute value of an edge is $\bar{c}$ then clearly the optimal solution of the traveling salesman problem is not smaller than $-\bar{c} \cdot n$ and not larger than $\bar{c} \cdot n$. Using the algorithm to solve the decision problem we can find the shortest tour length using the following approach.

procedure tsplength(G)

(1) Set $L = -\bar{c} \cdot n$ and $U = \bar{c} \cdot n$.

(2) As long as $L < U$ perform the following steps.

(2.1) Set $b = \lceil \frac{L+U}{2} \rceil$.

(2.2) If there exists a Hamiltonian tour of length at most b then set $U = b$, otherwise set $L = b + 1$.

end of tsplength

Applying this **binary search** technique we can find the length of the shortest tour by at most $\lceil \log(\bar{c} \cdot n) \rceil + 1$ calls of the solution algorithm for the TSP decision problem. (Throughout this text we will use log to denote the logarithm with base 2).
To completely solve the optimization problem we have to exhibit an optimal solution. This is now easily done once the shortest length is known.

procedure tsptour(G)

(1) Let U be the optimal tour length found by algorithm *tsplength*.

(2) For all $u = 1, 2, \ldots, n$ and all $v = 1, 2, \ldots, n$ perform the following steps.

(2.1) Set $s_{uv} = c_{uv}$ and $c_{uv} = \bar{c} \cdot n + 1$.

(2.2) If there does not exist a Hamiltonian tour of length U in the modified graph then restore $c_{uv} = s_{uv}$.

end of tsptour

After execution of this procedure the edges whose weights have not been altered give the edges of an optimal tour.
The procedures *tsplength* and *tsptour* call a polynomial number (in n and $\log \bar{c}$) of times the algorithm for the solution of the traveling salesman decision problem. Optimization problems with the property that they can be polynomially reduced to a decision problem in $\mathcal{NP}$ are called **$\mathcal{NP}$-easy**. Problems which are both $\mathcal{NP}$-easy and $\mathcal{NP}$-hard (like the traveling salesman problem) are called **$\mathcal{NP}$-equivalent**. If $\mathcal{P} \neq \mathcal{NP}$ then no $\mathcal{NP}$-hard problem can be solved in polynomial time, if $\mathcal{P}=\mathcal{NP}$ then every $\mathcal{NP}$-easy problem is in $\mathcal{P}$.
So far we have considered the general traveling salesman problem. One might hope that there are special cases where the problem can be solved in polynomial time. Unfortunately, such cases rarely have practical importance (BURKARD (1990), VAN DAL (1992), VAN DER VEEN (1992), WARREN (1993)). For most practical situations, namely for symmetric distances with triangle inequality, for Euclidean instances, for bipartite planar graphs, or even for grid graphs, the traveling salesman problem remains $\mathcal{NP}$-hard.
A different important issue is the question of whether algorithms can be designed which deliver solutions with requested or at least guaranteed quality in polynomial time (polynomial in the problem size and in the desired accuracy). Whereas for other $\mathcal{NP}$-hard problems such possibilities do exist, there are only discouraging results for the general TSP. For a particular problem instance let c_{opt} denote the length of a shortest tour and c_H denote the length of a tour computed by heuristic H. There are two basic results relating these two values.

Theorem 2.2 *Unless $\mathcal{P}=\mathcal{NP}$ there does not exist for any constant $r \geq 1$ a polynomial time heuristic H such that $c_H \leq r \cdot c_{\text{opt}}$ for all problem instances.*

□

A proof is given in SAHNI & GONZALES (1976).
A **fully polynomial approximation scheme** for a minimization problem is a heuristic H which computes for a given problem instance and any $\varepsilon > 0$ a feasible solution satisfying $c_H \leq (1+\varepsilon) \cdot c_{\text{opt}}$ in time polynomial in the size of the instance and in ε^{-1}.
It is an easy exercise to prove that to require polynomiality also in the encoding length of ε is equivalent to require a polynomial algorithm for the exact solution. It is very unlikely that fully polynomial approximation schemes exist for the traveling salesman problem since the following result holds.

Theorem 2.3 *Unless $\mathcal{P}=\mathcal{NP}$ there does not exist a fully polynomial approximation scheme for the Euclidean traveling salesman problem.*

□

A proof can be found in JOHNSON & PAPADIMITRIOU (1985). The result holds in general for TSPs with triangle inequality.
Despite these theoretical results we can nevertheless design heuristics that determine good or very good tours in practice. The theorems tell us that for every heuristic there are however problem instances where it fails badly. There are a few approximation results for problems with triangle inequality which will be addressed in Chapter 6.
It should be pointed out that the complexity of an algorithm derived by theoretical analysis might be insufficient to predict its behaviour when applied to real-world instances of a problem. This is mainly due to the fact that only worst case analysis is

performed which may be different from average behaviour, and, in addition, polynomial algorithms can cause a large amount of CPU time if the polynomial is not of low degree. In fact, for practical applications only algorithms having running time at most $O(n^3)$ would be rated efficient, but even algorithms with running times as low as $O(n^2)$ may not be applicable in certain situations.
Another point is, that the proof of $\mathcal{NP}$-hardness of a problem does not imply the nonexistence of a reasonable algorithm for the solution of problem instances arising in practice. It is the aim of this study to show that in the case of the traveling salesman problem algorithms can be designed which are capable of finding good approximate solutions to even large sized real-world instances within moderate time limits.

2.3 Linear and Integer Programming

Linear and integer programming is not a central topic of this tract. However, at some points we will make references to concepts and results of linear and integer programming. We give a short survey on these. Highly recommendable references in this area are the prize-winning books SCHRIJVER (1986) and NEMHAUSER & WOLSEY (1988).
Let A be an $m \times n$-matrix (constraint matrix), b be an m-vector (right hand side) and c be an n-vector (objective function), where all entries of A, b, and c are rational numbers. Given these data the linear programming problem is defined as follows.

Linear Programming Problem
Find a vector x^* maximizing the objective function $c^T x$ over the set $\{x \in \mathbf{Q} \mid Ax \leq b\}$. □

A linear program may be given in various forms which can all be transformed to the above. For example we may have equality constraints, nonnegativity conditions for some variables, or the objective function is to be minimized.
In its general form a linear programming problem is given as

$$\begin{aligned} \max\ c^T x + d^T y & \\ Ax + By &\leq a \\ Cx + Dy &= b \\ x &\geq 0 \end{aligned}$$

with appropriately dimensioned matrices and vectors.
A fundamental concept of linear programming is **duality**. The **dual linear program** to the program given above (which is then called the **primal linear program**) is defined as

$$\begin{aligned} \min\ u^T a + v^T b & \\ u^T A + v^T C &\geq c^T \\ u^T B + v^T D &= d^T \\ u &\geq 0. \end{aligned}$$

It is easily verified that the dual of the dual problem is again the primal problem. One important aspect of the duality concept is stated in the following theorem.

Theorem 2.4 *Let P and D be a pair of dual linear programs as defined above. Suppose there exist vectors (x^*, y^*) and (u^*, v^*) satisfying the constraints of P, resp., D. Then we have*

(i) The objective function value of (x^*, y^*) (in problem P) is less than or equal to the objective function value of (u^*, v^*) (in problem D).

(ii) Both problems have optimal solutions and their objective function values are equal.

□

Duality exhibits further relations between primal and dual problem. But since they are not important in the sequel we omit them here. Note in particular, that the dual problem can be used to give bounds for the optimal value of the primal problem (and vice versa).

The first algorithm for solving linear programming problems was the **Simplex method** invented by DANTZIG (1963). Since then implementations of this method have been considerably improved. Today, even very large sized linear programming problems with several ten thousands of variables and constraints can be solved routinely in moderate CPU time (BIXBY (1994)). The running time of the Simplex method cannot be bounded by a polynomial, in fact there are examples where exponential running time is necessary to solve a problem.

However, the linear programming problem, i.e., the problem of maximizing a linear objective function subject to linear constraints is in $\mathcal{P}$. This was proved in the famous papers KHACHIAN (1979) (using the **Ellipsoid method**) and KARMARKAR (1984) (using an **Interior-point method**). Though both of these algorithms are polynomial, only interior point methods are competitive with the Simplex method (LUSTIG, MARSTEN & SHANNO (1994)).

These facts illustrate again that complexity analysis is in the first place only a theoretical tool to assess hardness of problems and running time of algorithms.

A step beyond polynomial solvability is taken if we require feasible solutions to have integral entries. The integer linear programming problem is defined as follows.

Integer Linear Programming Problem

Let A, b, and c be appropriately dimensioned with rational entries. Find a vector x^* maximizing the objective function $c^T x$ over the set $\{x \in \mathbf{Q} \mid Ax \leq b, x \text{ integer}\}$. □

This problem is $\mathcal{NP}$-complete and no duality results are available. We show that the traveling salesman problem can be formulated as an integer linear program.

To be able to apply methods of linear algebra to graph theory we associate vectors to edge sets in the following way. Let $G = (V, E)$ be a graph. If $|E| = m$ then we denote by $\mathbf{Q}^E$ the m-dimensional rational vector space where the components of the vectors $x \in \mathbf{Q}^E$ are indexed by the edges $uv \in E$. We denote a component by x_{uv} or x_e if $e = uv$.

The **incidence vector** $x^F \in \mathbf{Q}^E$ of an edge set $F \subseteq E$ is defined by setting $x^F_{uv} = 1$ if $uv \in F$ and by setting $x^F_{uv} = 0$ otherwise. Similarly, if we associate a variable x_{uv} to each edge uv we denote by $x(F)$ the formal sum of the variables belonging to the edges of F.

the TSP for the complete graph K_n with edge weights c_{uv}. With the
that $x_{uv} = 1$ if edge uv is contained in a tour and $x_{uv} = 0$ otherwise, the
a formulation of the TSP as an integer linear program.

$$\min \sum_{uv \in E} c_{uv} x_{uv}$$

$$\begin{aligned} x(\delta(v)) &= 2, && \text{for all } u \in V, \\ x(C) &\leq |C| - 1, && \text{for all cycles } C \subseteq E_n, |C| < n, \\ x_{uv} &\in \{0,1\}, && \text{for all } u, v \in V. \end{aligned}$$

The TSP can be successfully attacked within the framework of linear and integer programming for surprisingly large problem sizes. We will comment on this issue in Chapter 12.

2.4 Data Structures

In this section we discuss some data structures that are useful for implementing traveling salesman problem algorithms. They are all used in the software package TSPX (REINELT (1991b)) with which all experiments in this monograph were conducted. The exposition is based on the books TARJAN (1983) and CORMEN, LEISERSON & RIVEST (1989) on algorithms and data structures. A further reference on the foundations of algorithms is KNUTH (1973).

2.4.1 Binary Search Trees

A **rooted tree** is a connected acyclic graph with one distinguished node, the so-called **root** or **root node** of the tree. Therefore, if the graph has n nodes a tree consists of $n-1$ edges. The **depth** of the tree is the length (number of edges) of the longest path from the root to any other node. Every node v is connected to the root by a unique path. The length of this path is said to be the **depth** or the **level** of v. We say that a tree is **binary** if at most two edges are incident to the root node and at most three edges are incident to every other node.
If node u is the first node encountered when traversing the unique path from v to the root then u is called **father** of v, denoted by $f[v]$ in the following. By definition $f[r] = 0$ if r is the root node. If v is a node in a binary tree then there are at most two nodes with v as their father. These nodes are called **sons** of v, and we define one of them to be the **right son** and the other one to be the **left son** of v. Either of them may be missing. A node without sons is called **leaf** of the tree. To represent a binary tree we store for every node v its father $f[v]$, its right son $r[v]$ and its left son $l[v]$. If one of them is missing we assign 0 to the respective entry.
Figure 2.1 shows a binary tree with root 5 and leaves 2, 7, 3, and 9.

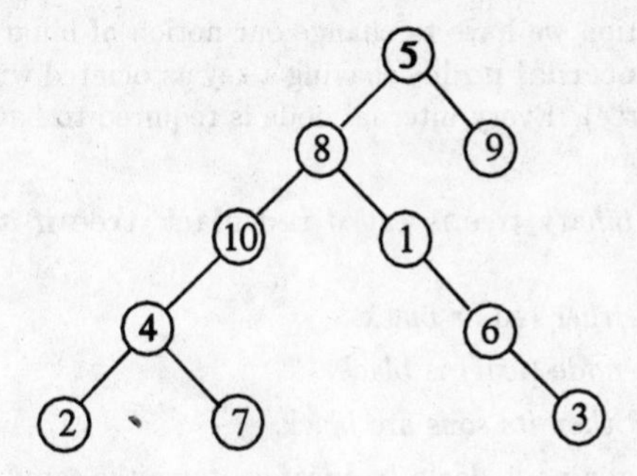

Figure 2.1 A binary tree on 10 nodes

By assigning to each node v a number $k[v]$, the **key** of the node, we can store information in such a tree.
Let $a_1, a_2, \ldots, a_n$ be a sequence of keys assigned to a set of n nodes.

Definition 2.5 *A binary tree on these nodes is called* **search tree** *if it satisfies the following conditions.*

(i) *If* $u = r[v]$ *then* $k[u] \geq k[v]$.
(ii) *If* $w = l[v]$ *then* $k[w] \leq k[v]$.

□

If the following procedure is called with the root of the search tree as parameter then it prints the stored numbers in increasing order.

procedure inorder(v)

(1) If $v = 0$ then return.

(2) Call *inorder*$(l[v])$.

(3) Print $k[v]$.

(4) Call *inorder*$(r[v])$.

end of inorder

Algorithms to find the smallest key stored in a binary search tree or to check if some key is present are obvious.
The depth of an arbitrary binary search tree can be as large as $n-1$ in which case the tree is a path. Hence checking if a key is present in a binary search tree can take time $O(n)$ in the worst case. This is also the worst case time for inserting a new key into the tree.
On the other hand, if we build the tree in a clever way, we could realize it with depth $\lfloor \log n \rfloor$. In such a tree searching can be performed much faster.
This observation leads us to the concept of **balanced binary search trees** which allow searching in worst case time $O(\log n)$. There are many possibilities for implementing balanced trees. We describe the so-called **red-black trees**.

For the proper definition we have to change our notion of binary search trees. Now we distinguish between **internal nodes** (having a key associated with them) and **external nodes** (leafs of the tree). Every internal node is required to have two sons, no internal node is a leaf of the tree.

Definition 2.6 *A binary tree is called* **red-black tree** *if it satisfies the following conditions.*

(i) Every node is either red or black.

(ii) Every external node (leaf) is black.

(iii) If a node is red then its sons are black.

(iv) Every path from a node down to a leaf contains the same number of black nodes.

□

It can be shown that a red-black tree with n non-leaf nodes has depth at most $2\log(n+1)$. Therefore searching in a red-black tree takes time $O(\log n)$.
Insertion of a new key can also be performed in time $O(\log n)$ because of the small depth of the tree. But we have to ensure that the red-black property still holds after having inserted a key. To do this we have to perform some additional fixing operations. Basically, we color the newly inserted node red and then reinstall the red-black condition on the path from the new node to the root. At every node on this path we spend constant time to check correctness and to fix node colors and keys if necessary. So the overall time spent for inserting a new node and reestablishing the red-black condition is $O(\log n)$.

2.4.2 Disjoint Sets Representation

Very frequently we need to manage a partition of some ground set $V = \{1, 2, \ldots, n\}$ into disjoint subsets $S_1, S_2, \ldots, S_k$, i.e., sets satisfying $\cup_{i=1}^{k} S_i = V$ and $S_i \cap S_j = \emptyset$ for all $i \neq j$. Operations to be performed are merging two subsets and identifying the subset containing a given element.
The data structure to store such a partition consists of a collection of trees each representing one subset. The nodes of each tree correspond to those elements of the ground set which belong to the respective subset, and the root node of each tree is said to be the **representative** of the respective set. For simplification of algorithms we define here the father of the root to be the root itself.
Suppose we have an initialized data structure representing some partition. Identification of the subset, i.e., the representative of the subset, to which an element $v \in V$ belongs is achieved by the following function.

function find(v)

(1) While $f[v] \neq v$ set $v = f[v]$.

(2) Return v.

end of find

Joining two subsets S and T where we are given elements $x \in S$ and $y \in T$ is accomplished by the following procedure.

procedure union(x, y)

(1) Set $u = \mathit{find}(x)$.

(2) Set $v = \mathit{find}(y)$.

(3) Set $f[u] = v$.

end of union

The representative of the new set is v.
We will exclusively use the disjoint set representation in the following context. The ground set V is the node set of some graph with edge set E. Initially, the ground set is partitioned into n sets containing one element each. We then scan the edges of E in some order depending on the application. If the endnodes of the current edge satisfy some condition then the sets containing these endnodes are merged. Usually $n-1$ merge operations are performed so that the final partition consists just of the set V. The number of find operations is bounded by $2|E|$. An application of this principle is used for implementing Kruskal's spanning tree algorithm to be discussed in section 2.5.
Without further modifications the above implementation can result in trees which are paths. This is the worst case for performing find operations. Ideally, we would like to have trees of depth 1 for set representation. But to achieve this we have to traverse one of the trees participating in a union operation. On the other hand, this can result in a running time of $O(m^2)$ for m union operations if the wrong trees are chosen.
Fortunately, there are two improvements to overcome these problems. We implement the find operations with additional **path compression** and union operations as **union by rank**. In addition we now store for each node r the number of nodes $n[r]$ in the tree rooted at r. The modified procedures are now.

function find_and_compress(v)

(1) If $v \neq f[v]$ then set $f[v] = \mathit{find\_and\_compress}(f[v])$.

(2) Return $f[v]$.

end of find_and_compress

After execution of this procedure for a node v all nodes on the path from v to the root (including v) in the previous tree now have the root node as their father.

procedure union_by_rank(x, y)

(1) Set $u = \mathit{find\_and\_compress}(x)$.

(2) Set $v = \mathit{find\_and\_compress}(y)$.

(3) If $n[u] < n[v]$ then set $f[u] = v$ and $n[v] = n[v] + n[u]$. Otherwise set $f[v] = u$ and $n[u] = n[u] + n[v]$.

end of union_by_rank

This procedure makes the root of the larger tree the father of the root of the smaller tree after the union operation. Looking more closely at this principle one realizes that it is not necessary to know the exact number of nodes in the trees. It suffices to store a rank at the root node which is incremented by one if trees of equal rank are merged. This way one can avoid additions of integers.
Initialization of a single element set is simply done by the following code.

procedure make_set(v)

(1) Set $f[v] = v$ and $n[v] = 1$.

end of make_set

The modified implementation of union/find turns out to perform very efficiently for our purposes.

Theorem 2.7 *If m operations are performed using disjoint sets representation by trees where n of them are make_set operations and the other ones are union operations (by rank) for disjoint sets and find operations (with path compression) then this can be performed in time $O(m \log^* n)$.*

□

A proof of this instructive theorem can be found in TARJAN (1983) or CORMEN, LEISERSON & RIVEST (1989). The number $\log^* n$ is defined via $\log^{(i)} n$ as follows.

$$\log^{(i)} n = \begin{cases} n & \text{if } i = 0, \\ \log(\log^{(i-1)} n) & \text{if } i > 0 \text{ and } \log^{(i-1)} n > 0, \\ \text{undefined} & \text{otherwise,} \end{cases}$$

and then

$$\log^* n = \min\{i \geq 0 \mid \log^{(i)} n \leq 1\}.$$

In fact, in the above theorem a slightly better bound of $O(m\alpha(m,n))$ where α denotes the inverse of the Ackermann function can be proved. But this is of no importance for practical computations since already $\log^* n \leq 5$ for $n \leq 2^{65536}$. So we can speak of linear running time of the fast union-find algorithm in practice (for our applications).

2.4.3 Heaps and Priority Queues

A **heap** is a data structure to store special binary trees satisfying an additional heap condition. These binary trees have the property that except for the deepest level all levels contain the maximal possible number of nodes. The deepest level is filled from "left to right" if we imagine the tree drawn in the plane. To every tree node there is an associated key, a real number. These keys are stored in a linear array A according to a special numbering that is assumed for the tree nodes. The root receives number 1 and if a node has number i then its left son has number $2i$ and its right son has number $2i+1$. If a node has number i then its key is stored in $A[i]$. Therefore, if such a binary tree has k nodes then the corresponding keys are stored in $A[1]$ through $A[k]$. The special property that array A has to have is the following.

Definition 2.8 *An array A of length n satisfies the* **heap property** *if for all $1 < i \leq n$ we have $A[\lfloor \frac{i}{2} \rfloor] \leq A[i]$.*

□

Stated in terms of binary trees this condition means that the key of the father of a node is not larger than the key of the node. The heap property implies that the root has the smallest key among all tree nodes, or equivalently, $A[1]$ is the smallest element of the array.
Alternatively, we can define the heap property as "$A[\lfloor \frac{i}{2} \rfloor] \geq A[i]$". This does not make an essential difference for the following, since only the relative order of the keys is reversed.
As a first basic operation we have to be able to turn an array filled with arbitrary keys into a heap. To do this we have to use the subroutine *heapify* with argument i which fixes the heap property for the subtree rooted at node i (where it is assumed that the two subtrees rooted at the left, resp. right, son of i are already heaps).

procedure heapify(i)

(1) Let n be the number of elements in heap A (nodes in the binary tree). If $2i > n$ or $2i+1 > n$ the array entries $A[2i]$, resp. $A[2i+1]$ are assumed to be $+\infty$.

(2) Let k be the index of i, $2i$, and $2i+1$ whose array entry is the smallest.

(3) If $k \neq i$ then exchange $A[i]$ and $A[k]$ and perform *heapify(k)*.

end of heapify

It is easy to see that *heapify(i)* takes time $O(h)$ if h is the length of the longest path from node i down to a leaf in the search tree. Therefore *heapify(1)* takes time $O(\log n)$. Suppose we are given an array A of length n to be turned into a heap. Since all leaves represent 1-element heaps the following procedure does the job.

procedure build_heap(A)

(1) For $i = \lfloor \frac{n}{2} \rfloor$ downto 1 perform *heapify(i)*.

end of build_heap

A careful analysis of this procedure shows that an arbitrary array of length n can be turned into a heap in time $O(n)$. Note that the binary tree represented by the heap is not necessarily a search tree.
Except for sorting (see section 2.5) we use the heap data structure for implementing **priority queues**. As the name suggests such a structure is a queue of elements such that elements can be accessed one after the other according to their priority. The first element of such a queue will always be the element with highest priority. For the following we assume that an element has higher priority than another element if its key is smaller.
The top element of a heap is therefore the element of highest priority. The most frequently applied operation on a priority queue is to extract the top-priority element and assure that afterwards the element of the remaining ones with highest priority is in the first position. This operation is implemented using the heap data structure. We assume that the current size of the heap is n.

function extract_top(A)

(1) Set $t = A[1]$.

(2) Set $A[1] = A[n]$ and decrease the heap size by 1.

(3) Call *heapify(1)*.

(4) Return t.

end of extract_top

Because of the call of *heapify(1)*, extracting the top element and fixing the heap property needs time $O(\log n)$. Inserting a new element is accomplished as follows.

procedure insert_key(k)

(1) Increase the heap size by 1 and set i to the new size.

(2) While $i > 1$ and $A[\lfloor \frac{i}{2} \rfloor] > k$

(2.1) Set $A[i] = A[\lfloor \frac{i}{2} \rfloor]$ and $i = \lfloor \frac{i}{2} \rfloor$.

(3) Set $A[i] = k$.

end of insert_key

Since in the worst case we have to scan the path from the new leaf to the root, a call of *insert_key* takes time $O(\log n)$.

2.4.4 Graph Data Structures

Very frequently we have to store undirected graphs $G = (V, E)$ with $|V| = n$ nodes and $|E| = m$ edges where n is large, say in the range of 1,000 to 100,000. The number of edges to be stored depends on the application.
Matrix type data structures are the (node-edge) incidence matrix and the (node-node) adjacency matrix. The **incidence matrix** A is an $n \times m$-matrix whose entries a_{ie} are defined by

$$a_{ie} = \begin{cases} 1 & \text{if } i \text{ is an endnode of edge } e, \\ 0 & \text{otherwise.} \end{cases}$$

The **adjacency matrix** is an $n \times n$-matrix B whose entries b_{ij} are defined by

$$b_{ij} = \begin{cases} 1 & \text{if } ij \text{ is an edge of } G, \\ 0 & \text{otherwise.} \end{cases}$$

Since we need $O(nm)$ or $O(n^2)$ storage for these matrices they cannot be used for large graphs. This also limits the use of distance matrices for the definition of edge weights in large problem instances. Fortunately, for large TSPs, distances between nodes are usually given by a distance function.
If we just want to store a graph we can use an **edge list** consisting of two arrays *tail* and *head* such that *tail*[e] and *head*[e] give the two endnodes of the edge e. This is

appropriate if some graph is generated edge by edge and no specific operation has to be performed on it.

Another possibility is to use a system of **adjacency lists**. Here we store for each node a list of its adjacent nodes. This is done by an array *adj* of length $2m$ containing the adjacent nodes and an array *ap* of length n. These arrays are initialized such that the neighbors of node i are given in *adj*[*ap*[i]], *adj*[*ap*[i] + 1], through *adj*[*ap*[$i+1$] − 1]. This data structure is suitable if we have to scan neighbors of a node and if the graph remains unchanged. Adding edges is time consuming since large parts of the array may have to be moved.

If we have to add edges and want to avoid moving parts of arrays, then we have to use linked lists. Since this is our most frequent operation on graphs we have used the following data structure to store an undirected graph. The two arrays *tail* and *head* contain the endnodes of the edges. The arrays *nxtt* and *nxth* are initialized such that *nxtt*[*e*] gives the number of a further edge in this structure having *tail*[*e*] as one endnode and *nxth*[*e*] gives the number of a further edge having *head*[*e*] as an endnode. An entry 0 terminates a linked list of edges for a node. For each node v the array entry *first*[v] gives the number of the first edge having v as an endnode.

To get used to this form of storing a graph we give an example here. Suppose we have a subgraph of the complete graph on six nodes consisting of the edges $\{1,2\}$, $\{1,5\}$, $\{2,5\}$, $\{2,3\}$, $\{3,5\}$, $\{4,5\}$ and $\{4,6\}$. This graph could be stored e.g., by the assignment shown in Table 2.2.

Index	*first*
1	6
2	5
3	5
4	7
5	7
6	4

Index	*head*	*tail*	*nxth*	*nxtt*
1	1	2	0	0
2	3	5	0	0
3	2	5	1	2
4	4	6	0	0
5	2	3	3	2
6	1	5	1	3
7	4	5	4	6

Table 2.2 Example for subgraph data structure

Suppose the current graph has m edges and is stored using this data structure. Adding a new edge can then be performed in constant time using the following piece of code.

procedure add_edge(i, j)

(1) Set *tail*[$m+1$]=i, *nxtt*[$m+1$]=*first*[i], and *first*[i] $= m+1$.

(2) Set *head*[$m+1$]=j, *nxth*[$m+1$]=*first*[j], and *first*[j] $= m+1$.

(3) Set $m = m+1$.

end of add_edge

Of course, if we add more edges than dynamic memory space has been allocated for we have to allocate additional space.

2.4.5 Representing Tours

An easy way to store a tour is to use an array t of length n and let $t[k]$ be the k-th node visited in the tour. However, this is not sufficient as we shall see below. We also have to impose a direction on the tour. We therefore store with each node i its predecessor *pred*[i] and its successor *succ*[i] in the tour with respect to the chosen orientation.

When using heuristics to find short tours one has to perform a sequence of local modifications of the current tour to improve its length. We explain our method to perform modifications in an efficient way using the example of 2-opt moves. A **2-opt move** consists of removing two edges from the current tour and reconnecting the resulting paths in the best possible way. This operation is depicted in Figure 2.3 where broken arcs correspond to directed paths.

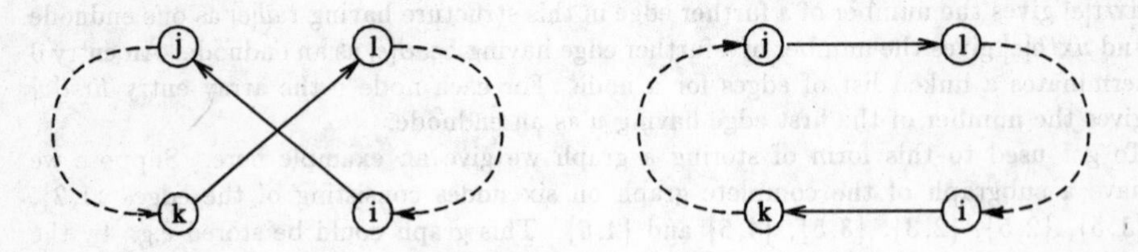

Figure 2.3 A 2-opt move

Note that we have to have an imposed direction of the tour to be able to decide which pair of new edges can be added to form a new tour. Adding edges jk and il would result in an invalid subgraph consisting of two subtours. Furthermore, the direction on one of the paths has to be reversed which takes time $O(n)$.

Since we have to make a sequence of such 2-opt moves we do not update the tour structure as in Figure 2.3 but rather store the current tour as a sequence of unchanged intervals of the starting tour. This is of particular importance for some heuristics where 2-opt moves are only tentative and might not be realized to form the next tour. For each interval we store the direction in which it is traversed in the current tour. The result of the 2-opt move in our example would be a doubly linked sequence of intervals as in Figure 2.4 where we also indicate that the path represented by the interval $[k, j]$ has been reversed.

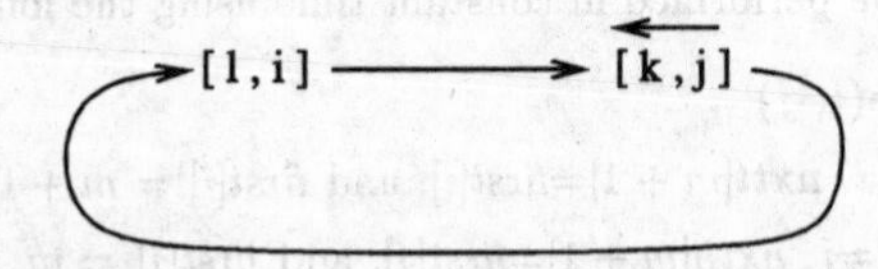

Figure 2.4 Result of the 2-opt move

To make the approach clearer we use concrete node numbers $l = 4$, $i = 7$, $k = 3$, $j = 10$ and perform another 2-opt move involving nodes 11 and 5 on the path from 4 to 7 and nodes 8 and 16 on the path from 3 to 10 as in Figure 2.5.

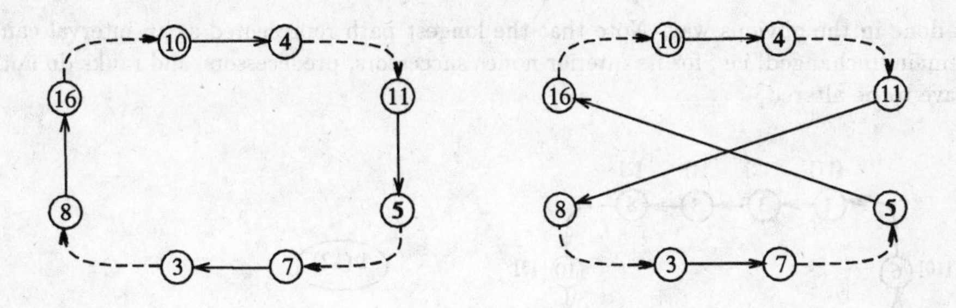

Figure 2.5 A second 2-opt move

After execution of this move we have the interval sequence shown in Figure 2.6. Note that the segment between nodes 3 and 8 was reversed before the 2-opt move, so it is not reversed any more after the move.

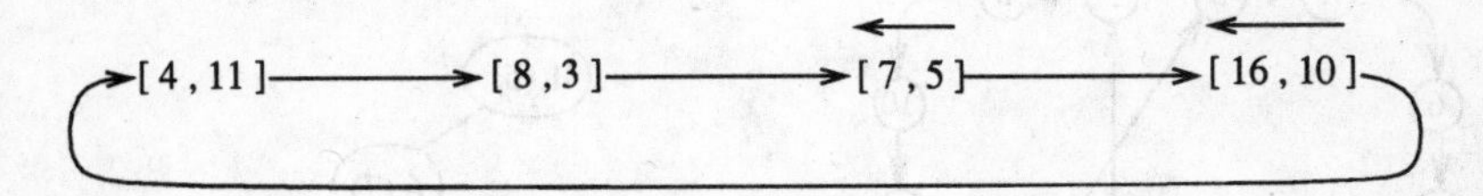

Figure 2.6 Result of the second 2-opt move

The new interval sequence was obtained by splitting two intervals and reconnecting the intervals in an appropriate way. Note that also in the interval representation of a tour we have to reorient paths of the sequence. But, if we limit the number of moves in such a sequence by k this can be performed in $O(k)$.

One difficulty has been omitted so far. If we want to perform a 2-opt move involving four nodes we have to know the intervals in which they are contained to be able to choose the correct links. We cannot do this efficiently without additional information. We store with each node its **rank** in the starting tour. This rank is defined as follows: An arbitrary node gets rank 1, its successor gets rank 2, etc., until the predecessor of the rank 1 node receives rank n. Since we know for each interval the ranks of its endnodes and whether it has been reversed with respect to the starting tour or not, we can check which interval contains a given node if we store the intervals in a balanced binary search tree.

Applying this technique the interval containing a given node can be identified in time $O(\log k)$ if we have k intervals.

This way of maintaining tour modifications is extensively used in the implementation of a Lin-Kernighan type improvement method (see Chapter 7). Experience with this data structure was also reported in APPLEGATE, CHVÁTAL & COOK (1990). They used splay trees (TARJAN (1983)) instead of red-black trees in their implementation.

Of course, the number of intervals should not become too large because the savings in execution time decreases with the number of intervals.

Finally, or if we have too many intervals, we have to clear the interval structure and generate correct successor and predecessor pointers to represent the current tour. This is done in the obvious way. Note that the longest path represented as an interval can remain unchanged, i.e., for its interior nodes successors, predecessors, and ranks do not have to be altered.

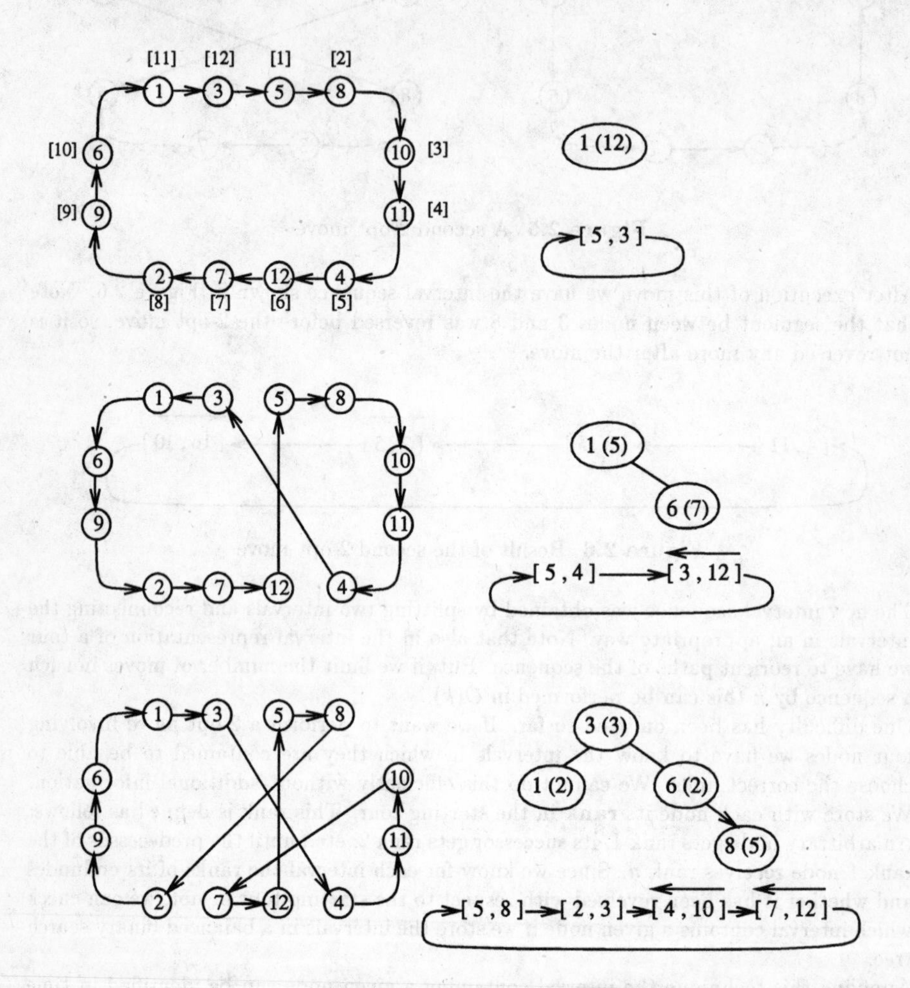

Figure 2.7 An example for representing a tour

To visualize this approach we give a sequence of 2-opt moves starting from a basic tour together with the resulting interval sequences and search trees in Figure 2.7. Ranks of nodes are listed in brackets. Each node of the search tree represents an interval. We give for each node the rank of the endnode with lower rank and in parentheses the number of nodes in the interval.

2.5 Some Fundamental Algorithms

In this chapter we review some basic algorithms which are not traveling salesman problem specific but are used as building blocks in many heuristics. For an extensive discussion we refer again to CORMEN, LEISERSON & RIVEST (1989).

2.5.1 Sorting

Given a set $A = \{a_1, a_2, \ldots, a_n\}$ of n integer or rational numbers, the sorting problem consists of finding the sequence of these numbers in increasing (or decreasing) order. In many cases the numbers will correspond to the weights of the edges of a subgraph.
There is a variety of algorithms we cannot discuss here. One example of a very simple sorting algorithm shall be given first. This algorithm recursively subdivides a set into two halves, sorts the subsets and then merges the sorted sequences. Suppose the numbers are stored in $B[1]$, $B[2]$, through $B[n]$.

procedure mergesort(B, l, u)

(1) If $l \geq u - 1$ sort $B[l]$ through $B[u]$ by comparisons and return.

(2) Perform *mergesort*$(B, l, \lfloor \frac{l+u}{2} \rfloor)$.

(3) Perform *mergesort*$(B, \lfloor \frac{l+u}{2} \rfloor + 1, u)$.

(4) Rearrange B to represent the sorted sequence.

end of mergesort

The call *mergesort*$(B, 1, n)$ sorts the array in time $O(n \log n)$. This will follow from considerations in section 2.5.3 since Step (1) is executed in constant time and Step (4) can be performed in time $O(u - l)$.
Another sorting algorithm makes use of the heap data structure presented in the previous chapter. Since in a heap the top element is always the smallest one we can successively generate the sorted sequence of the elements of A using the heap.

procedure heapsort(B)

(1) Call *build_heap*(B).

(2) For $i = n$ downto 1 perform the following steps.

(2.1) Exchange $B[1]$ and $B[i]$.

(2.2) Decrement the heap size by 1.

(2.3) Call *heapify*(1).

end of heapsort

After execution of *heapsort(A)* we have the elements of A sorted in increasing order in $A[n]$, $A[n-1]$, through $A[1]$.
The running time is easily derived from the discussion in the section on heaps. Step (1) takes time $O(n)$, and, since Step (2.3) takes time $O(\log i)$, we obtain the overall running time as $O(n) + \sum_{i=1}^{n} O(\log i) = O(n \log n)$.
Therefore both merge sort and heap sort seem to be more or less equivalent. However, heap sort has an important advantage. It is able to generate the sorted sequence as long as needed. If for some reason the remaining sequence is not of interest any more at some point we can exit from heapsort prematurely.
A final remark is in order. It can be shown that sorting based on the comparison of two elements cannot be performed faster than in $O(n \log n)$ time. So the above discussion shows that the sorting problem has time complexity $\Theta(n \log n)$ (in this computational model).
Faster sorting algorithms can only be achieved if some assumptions on the input can be made, e.g., that all numbers are integers between 1 and n. Expected linear running time of some sorting algorithms can be shown for specific input distributions.

2.5.2 Median Finding

We could also have implemented a sorting algorithm by recursively doing the following for a set $A = \{a_1, a_2, \ldots, a_n\}$. First identify a **median** of A, i.e., a value $\overline{a}$ such that half of the elements of A are below $\overline{a}$ and half of the elements are above $\overline{a}$. More precisely, we identify a number $\overline{a}$ such that we can partition A into two sets A_1 and A_2 satisfying $A_1 \subseteq \{a_i \mid a_i \leq \overline{a}\}$, $A_2 \subseteq \{a_i \mid a_i \geq \overline{a}\}$, $|A_1| = \lfloor \frac{n}{2} \rfloor$, and $|A_2| = \lceil \frac{n}{2} \rceil$. We then sort A_1 and A_2 separately. The concatenation of the respective sorted sequences give a sorting of A.
In particular for geometric problems defined on points in the plane we often need to compute horizontal or vertical lines separating the point set into two (approximately equally sized) parts. For this we need medians with respect to the x- or y-coordinates of the points.
A natural way to find a median is to sort the n points. The element at position $\lfloor \frac{n}{2} \rfloor$ gives a median. However, one can do better as is shown in the following sophisticated algorithm which is also very instructive. The algorithm requires a subroutine that for a given input b rearranges an array such that in the first part all elements are at most as large as b while in the second part all elements are at least as large as b. Assume again that array B contains n numbers in $B[1]$ through $B[n]$.

function partition(B, b)

(1) Set $i = 1$ and $j = n$.

(2) Repeat the following steps until $i \geq j$.

(2.1) Decrement j by 1 until $B[j] \leq b$.

(2.2) Increment i by 1 until $B[i] \geq b$.

(2.3) If $i < j$ exchange $B[i]$ and $B[j]$.

(3) Return j.

end of partition

After execution of this algorithm we have $B[l] \leq b$ for all $l = 1, 2, \ldots, i$ and $B[l] \geq b$ for all $l = j, j+1, \ldots, n$.
The procedure for finding a median can now be given. In fact, the procedure does a little bit more. It finds the i-th smallest element of a set.

procedure find_ith_element(B, i)

(1) Partition B into $\lfloor \frac{n}{5} \rfloor$ groups of 5 elements each and one last group containing the remaining elements.

(2) Sort each set to find its "middle" element. If the last set has even cardinality l we take the element at position $\frac{l}{2} + 1$.

(3) Apply *find_ith_element* to find the median b of the set of medians found in Step (2).

(4) Let $k = \mathit{partition}(B, b)$.

(5) If $i \leq k$ then find the i-th smallest element of the lower side of the partition, otherwise find the $(i - k)$-th smallest element of the higher side of the partition.

end of find_ith_element

The call *find_ith_element*$(B, \lfloor \frac{n}{2} \rfloor)$ now determines a median of B.
A running time analysis shows that this algorithm runs in linear time which is best possible since every element of B has to be considered in a median finding procedure. Hence medians can be found in time $\Theta(n)$.

2.5.3 Divide and Conquer

We have already applied the divide and conquer principle without having named it explicitly. The basic idea is to divide a problem into two (or more) subproblems, solve the subproblems, and then construct a solution of the original problem by combining the solutions of the subproblems in an appropriate way.
Since this is a very important and powerful principle, we want to cite the main result for deriving the running time of a divide and conquer algorithm of CORMEN, LEISERSON & RIVEST (1989). Such an algorithm has the following structure where we assume that it is applied to some set S, $|S| = n$, to solve some (unspecified) problem.

procedure divide_and_conquer(S)

(1) Partition S into subproblems $S_1, S_2, \ldots, S_a$ of size less than $\frac{n}{b}$.

(2) For each subproblem S_i perform *divide_and_conquer*(S_i).

(3) Combine the subproblem solutions to solve the problem for S.

end of divide_and_conquer

In the following we assume that f is the running time function for performing Steps (1) and (3).
Theorem 2.9 gives a formula to compute the running time of divide and conquer algorithms. (It does not matter that $\frac{n}{b}$ is not necessarily integral, we can both substitute $\lfloor\frac{n}{b}\rfloor$ or $\lceil\frac{n}{b}\rceil$).

Theorem 2.9 *Let $a \geq 1$ and $b > 1$ be constants and let $T(n)$ be defined by the recurrence $T(n) = aT(\frac{n}{b}) + f(n)$. Then $T(n)$ can be bounded asymptotically (depending on f) as follows.*

(i) If $f(n) = \Theta(n^{\log_b a - \varepsilon})$ for some constant $\varepsilon > 0$, then $T(n) = \Theta(n^{\log_b a})$.

(ii) If $f(n) = O(n^{\log_b a})$, then $T(n) = \Theta(n^{\log_b a} \log n)$.

(iii) If $f(n) = \Omega(n^{\log_b a + \varepsilon})$ for some constant $\varepsilon > 0$ and if $a \cdot f(\frac{n}{b}) \leq c \cdot f(n)$ for some constant $c < 1$ and all sufficiently large n, then $T(n) = \Theta(f(n))$.

□

This theorem provides a very useful tool. Take as an example the merge sort algorithm. Step (1) is performed in constant time and Step (4) is performed in time $O(u-l)$, hence $f(n) = O(n)$. For the theorem we have $a = 2$ and $b = 2$. Therefore case (ii) applies and we obtain a running time of $\Theta(n \log n)$ for merge sort.

2.5.4 Minimum Spanning Trees

Let $G = (V, E)$, $|V| = n$, $|E| = m$, be a connected graph with edge weights c_{uv} for all $uv \in E$. A **minimum spanning tree** of G is an acyclic connected subset $T \subseteq E$ such that $c(T)$ is minimal among these edge sets. Clearly a spanning tree has $n-1$ edges. The following algorithm computes a minimum length spanning tree of G (PRIM (1957)).

procedure prim(G)

(1) Set $T = \emptyset$ and $Q = \{1\}$.

(2) For all $i = 2, 3, \ldots, n$ set $d[i] = c_{1i}$ and $p[i] = 1$ if $\{1, i\} \in E$, resp., $d[i] = \infty$ and $p[i] = 0$ if $\{1, i\} \notin E$.

(3) As long as $|T| < n-1$ perform the following.

(3.1) Let $d[j] = \min\{d[l] \mid l \in V \setminus Q\}$.

(3.2) Add edge $\{j, p[j]\}$ to T and set $Q = Q \cup \{j\}$.

(3.3) For all $l \in V \setminus Q$ check if $c_{jl} < d[l]$. If this is the case set $d[l] = c_{jl}$ and $p[l] = j$.

(4) T is a minimum spanning tree of G.

end of prim

The running time of this algorithm is $\Theta(n^2)$. It computes in each iteration a minimum spanning tree for the subgraph $G_Q = (Q, E(Q))$, and so upon termination of the algorithm we have a minimum spanning tree of G.
The implementation of Prim's algorithm given here is best possible if we have complete or almost complete graphs G. If we want to compute minimum spanning trees for graphs with fewer edges then other implementations are superior. The idea is to maintain the nodes not yet connected to the tree in a binary heap (where the keys are the shortest distances to the tree). Since keys are only decreased, every update in Step (3.3) can be performed in time $O(\log n)$. Because we have to scan the adjacency list of j to see which distances might be updated, Step (3.3) takes altogether time $O(m \log n)$. Finding the minimum of a heap and updating it has to be performed $n-1$ times requiring time $O(\log n)$ each time. Therefore we obtain running time $O(m \log n)$ for the heap implementation of Prim's algorithm.
Using more advanced techniques (as e.g., binomial heaps or Fibonacci heaps) one can even achieve time $O(m + n \log n)$.
A different approach to finding a minimum spanning tree was given in KRUSKAL (1956). It is particularly suited for sparse graphs.

procedure kruskal(G)

(1) Build a heap of the edges of G (with respect to smaller weights).

(2) Set $T = \emptyset$.

(3) As long as $|T| < n-1$ perform the following.

(3.1) Get the top edge $\{u, v\}$ from the heap and update the heap.

(3.2) If u and v belong to different connected components of the graph (V, T) then add edge $\{u, v\}$ to T.

(4) T is a minimum spanning tree of G.

end of kruskal

The idea of this algorithm is to generate minimum weight acyclic edge sets with increasing cardinality until such an edge set of cardinality $n-1$ is found (which is then a minimum spanning tree of G).
This algorithm is implemented using fast union-find techniques. Combining results for maintaining heaps and for applying fast union-find operations we obtain a running time of $O(m \log m)$ for this algorithm.
Instead of using a heap we could sort in a first step all edges with respect to increasing length and then use only union-find to identify different components. Using a heap we can stop as soon as $|T| = n-1$ and do not necessarily have to sort all edges.
If G is not connected a slight modification of the above computes the minimum weight acyclic subgraph of cardinality $n-k$ where k is the number of connected components of G.

2.5.5 Greedy Algorithms

The **greedy algorithm** is a general algorithmic principle for finding feasible solutions of combinatorial optimization problems. It is characterized by a myopic view, performing the construction of a feasible solution step by step based only on local knowledge of the problem.
To state it in general, we need a suitable definition of a combinatorial optimization problem. Let $E = \{e_1, e_2, \ldots, e_m\}$ be a finite set where each element has an associated weight $c_i, 1 \leq i \leq m$, and $\mathcal{I} \subseteq 2^E$ be the set of so-called feasible solutions. For a set $I \subseteq E$, its weight is given as $c(I) = \sum_{e_i \in I} c_i$. The optimization problem consists of finding a feasible solution $I^* \in \mathcal{I}$ such that $c(I^*) = \min\{c(I) \mid I \in \mathcal{I}\}$.
The greedy algorithm works as follows.

procedure greedy($\mathcal{I}$, c)

(1) Sort E such that $c_1 \leq c_2 \leq \ldots, c_m$.

(2) Set $I = \emptyset$.

(3) For $i = 1, 2, \ldots, m$:

(3.1) If $I \cup \{e_i\} \in \mathcal{I}$, then set $I = I \cup \{e_i\}$.

end of greedy

Due to the sorting in Step (1), the algorithm needs time $\Omega(m \log m)$. It does not need more time, if the test "$I \cup \{e_i\} \in \mathcal{I}$?" can be performed in time $O(\log m)$.
Inspite of its simplicity, the greedy principle has some important aspects, in particular in the context of matroids and independence systems, but we do not elaborate on this. In general, however, it delivers feasible solutions of only moderate quality. Note, that Kruskal's algorithm for computing minimum spanning trees is nothing but the greedy algorithm. Because the spanning tree problem is essentially an optimization problem over a matroid, the greedy algorithm is guaranteed to find an optimal solution in this case.

Chapter 3

Related Problems and Applications

For several practical problems it is immediately seen that the TSP provides the suitable optimization model. In many cases, however, this is either not straightforward or the pure TSP has to be augmented by further constraints. In this chapter we first discuss some optimization problems that are related to the TSP. Some of them can be transformed to a pure TSP in a reasonable way, others are at least related in the sense that algorithms developed for the TSP can be adapted to their solution. Then we survey some application areas where the TSP or its relatives can be used to treat practical problems. Further aspects are found in GARFINKEL (1985). Finally, we introduce the collection of sample problem instances that we will use in the sequel for testing algorithms.

3.1 Some Related Problems

Note that we can assume without loss of generality that a symmetric TSP is always a minimization problem and that all distances c_{ij} are positive. First, if we are looking for the longest Hamiltonian cycle we can multiply all edge weights by -1 and solve a minimization problem. Second, we can add a constant to all edge weights without affecting the ranking of tours with respect to their lengths. Hence all edge weights can be made positive. In the sequel we will not explicitly mention this fact. Without loss of generality we can also assume that all edge lengths are integer numbers.
Observe, however, that approximation results may be no longer valid after having modified edge weights. For example, if a very large constant is added to each edge weight, then every tour is near optimal.

3.1.1 Traveling Salesman Problems in General Graphs

There may be situations where we want to find shortest Hamiltonian tours in arbitrary graphs $G = (V, E)$, in particular in graphs which are not complete. If it is required that each node is visited exactly once and that only edges of the given graph must be used then we can do the following. Add all missing edges giving them a sufficiently large weight M (e.g., $M = \sum_{ij \in E} c_{ij}$) and apply an algorithm for the symmetric TSP in complete graphs. If this algorithm terminates with an optimal tour containing none of the edges with weight M, then this tour is also optimal for the original problem. If an edge with weight M is contained in the optimal tour then the original graph does not

contain a Hamiltonian cycle. Note however, that heuristics cannot guarantee to find a tour in G even if one exists.
The second way to treat such problems is to allow that nodes may be visited more than once and edges be traversed more than once. If the given graph is connected we can always find a feasible roundtrip under this relaxation. This leads us to the so-called graphical traveling salesman problem.

3.1.2 The Graphical Traveling Salesman Problem

For an arbitrary connected graph G with edge weights, the **graphical traveling salesman problem (GTSP)** consists of finding a closed walk in G for the salesman to visit every city requiring the least possible total distance. The salesman may only use edges of G, but is allowed to visit a city or to traverse an edge more than once. This is sometimes a more practical definition of the TSP because we may have cases where the underlying graph of connections does not even contain a Hamiltonian cycle and where some direct transitions from a city i to a city j are not possible. In the formulation as a GTSP we explicitly stick to the given graph.
To avoid degenerate situations we have to have nonnegative edge weights. Otherwise we could use an edge as often as we like in both directions to achieve an arbitrarily small length of the solution. We transform a GTSP to a symmetric TSP as follows. Let $K_n = (V_n, E_n)$ be the complete graph on n nodes. For every pair i, j of nodes we compute the shortest path from i to j in the graph G. The length d_{ij} of this path is taken as the weight of edge ij in K_n. Now the shortest Hamiltonian tour in K_n can be transformed to a shortest closed walk in G visiting all nodes.
NADDEF & RINALDI (1993) discuss relation between the TSP and the GTSP in detail.

3.1.3 The Shortest Hamiltonian Path Problem

We are given a graph $G = (V, E)$ with edge weights c_{ij}. Two special nodes, say v_s and v_t, of V are also given. The task is to find a path from v_s to v_t visiting each node of V exactly once with minimal length, i.e., to find the shortest Hamiltonian path in G from v_s to v_t.
This problem can be solved as a standard TSP in two ways.

a) Choose M sufficiently large and assign weight $-M$ to the edge from v_s to v_t. Then compute the optimal traveling salesman tour in this graph. This tour has to contain edge $v_s v_t$ and thus solves the Hamiltonian path problem.

b) Add a new node 0 to V and edges from 0 to v_s and to v_t with weight 0. Each Hamiltonian tour in this new graph corresponds to a Hamiltonian path from v_s to v_t in the original graph with the same length.

If the terminating point of the Hamiltonian path is not fixed, then we can solve the problem by introducing a new node 0 and adding edges from all nodes $v \in V \setminus \{v_s\}$ to 0 with zero length. Now we can solve the Hamiltonian path problem with starting point v_s and terminating point $v_t = 0$ which solves the original problem.
If neither starting point nor terminating point are specified, then we just add node 0 and connect all other nodes to 0 with edges of length zero. In this new graph we solve the standard TSP.

3.1.4 Hamiltonian Path and Cycle Problems

Sometimes it has to be checked if a given graph $G = (V, E)$ contains a Hamiltonian cycle or path at all. This question can be answered by solving a symmetric TSP in the complete graph K_n (with $n = |V|$) where all edges of the original graph obtain weight 1 and all other edges obtain weight 2. Then G contains a Hamiltonian cycle if and only if the shortest Hamiltonian cycle in K_n has length n. If this shortest cycle has length $n + 1$, then G is not Hamiltonian, but contains a Hamiltonian path.

3.1.5 The Asymmetric Traveling Salesman Problem

If the cost of traveling from city i to city j is not necessarily the same as of traveling from city j to city i, then an asymmetric traveling salesman problem has to be solved. Let $D = (W, A)$, $W = \{1, 2, \ldots, n\}$, $A \subseteq W \times W$, and d_{ij} be the arc weight of $(i, j) \in A$. We define a graph $G = (V, E)$ by

$$\begin{aligned} V &= W \cup \{n+1, n+2, \ldots, 2n\}, \\ E &= \{(i, n+i) \mid i = 1, 2, \ldots, n\} \\ &\quad \cup \{(n+i, j) \mid (i, j) \in A\}. \end{aligned}$$

Edge weights are assigned as follows

$$\begin{aligned} c_{i,n+i} &= -M \text{ for } i = 1, 2, \ldots, n, \\ c_{n+i,j} &= d_{ij} \quad \text{for } (i, j) \in A, \end{aligned}$$

where M is a sufficiently large number, e.g., $M = \sum_{(i,j) \in A} d_{ij}$. It is easy to see that for each directed Hamiltonian cycle in D with length d_D there is a Hamiltonian cycle in G with length $c_G = d_D - nM$. In addition, since an optimal tour in G contains all edges with weight $-M$, it induces a directed Hamiltonian cycle in D. Hence we can solve asymmetric TSPs as symmetric TSPs.

3.1.6 The Multisalesmen Problem

We give the asymmetric version of this problem. Instead of just one salesman now there are m salesmen available who are all located in a city $n + 1$ and have to visit cities $1, 2, \ldots, n$. The task is to select some (or all) of these salesmen and assign tours to them such that in the collection of all these tours together each city is visited exactly once. The activation of salesman j incurs a fixed cost w_j. The cost of the tour of salesman j is the sum of the intercity distances of his tour (starting at and returning to city $n+1$). The **m-salesmen problem (m-TSP)** now consists of selecting a subset of the salesmen and assigning a tour to each of them such that each city is visited by exactly one salesman and such that the total cost of visiting all cities this way is as small as possible.

In BELLMORE & HONG (1974) it was observed that this problem can be transformed to an asymmetric TSP involving only one salesman. We give their construction.

Let $D = (V, A)$ be the digraph where $V = \{1, 2, \ldots, n, n+1\}$ and $A \subseteq V \times V$ gives the possible transitions between cities. Let d_{ij} be the distance from city i to city j.
We construct a new digraph $D = (V', A')$ as follows.

$$\begin{aligned} V' &= V \cup \{n+2, n+3, \ldots, n+m\}, \\ A' &= A \\ &\quad \cup \{(n+i, j) \mid 2 \leq i \leq m, (n+1, j) \in A\} \\ &\quad \cup \{(j, n+i) \mid 2 \leq i \leq m, (j, n+1) \in A\} \\ &\quad \cup \{(n+i, n+i-1) \mid 2 \leq i \leq m\}. \end{aligned}$$

The weights d'_{ij} for the arcs (i, j) in A' are defined via

$$\begin{aligned} d'_{ij} &= d_{ij}, \text{ for } 1 \leq i \leq n, 1 \leq j \leq n, (i,j) \in A, \\ d'_{n+i,j} &= d_{n+1,j} + \tfrac{1}{2} w_i, \text{ for } 1 \leq i \leq m, 1 \leq j \leq n, (n+1, j) \in A, \\ d'_{j,n+i} &= d_{j,n+1} + \tfrac{1}{2} w_i, \text{ for } 1 \leq i \leq m, 1 \leq j \leq n, (j, n+1) \in A, \\ d'_{n+i,n+i-1} &= \tfrac{1}{2} w_{i-1} - \tfrac{1}{2} w_i, \text{ for } 2 \leq i \leq m. \end{aligned}$$

It is not difficult to verify that the shortest tour in D' relates to an optimal solution of the corresponding m-salesmen problem for D.
Observe in addition, that with an easy modification we can require that every salesman is to be activated. We simply eliminate all edges $(n+i, n+i-1)$ for $2 \leq i \leq m$. Of course, the fixed costs w_j can now be ignored.
A different transformation is given in JONKER & VOLGENANT (1988). Solution algorithms are discussed in GAVISH & SRIKANTH (1986).

3.1.7 The Rural Postman Problem

We are given a graph $G = (V, E)$ with edge weights c_{ij} and a subset $F \subseteq E$. The **Rural Postman Problem** consists of finding a shortest closed walk in G containing all edges in F. If $F = E$ we have the special case of a **Chinese postman problem** which can be solved in polynomial time using matching techniques (EDMONDS & JOHNSON (1973)). The standard symmetric TSP can easily be transformed to a rural postman problem. Therefore, in general the rural postman problem is $\mathcal{NP}$-hard. In Chapter 11 we will encounter a special version of the rural postman problem and find approximative solutions for it using TSP methods.

3.1.8 The Bottleneck Traveling Salesman Problem

Instead of tours with minimal total length one searches in this problem for tours whose longest edge is as short as possible. This **bottleneck traveling salesman problem** can be solved by a sequence of TSPs. To see this observe that the absolute values of the distances are not of interest under this objective function. We may reduce distances as long as they compare exactly as before. Hence we may assume that we have at most $\frac{1}{2}n(n-1)$ different distances and that the largest of them is not greater than $\frac{1}{2}n(n-1)$. We now solve problems of the following kind for some parameter b.

"Is there a Hamiltonian cycle of the original graph consisting only of edges with length at most b?"

This problem can be transformed to a standard TSP. By performing a binary search on the parameter b (starting with $b = \frac{1}{4}n(n-1)$) we can identify the smallest such b leading to a "yes" answer by solving at most $O(\log n)$ TSPs.

3.1.9 The Prize Collecting Traveling Salesman Problem

We are given a graph $G = (V, E)$ with edge weights c_{ij}, node weights u_i (representing benefits received when visiting the respective city), and a special base node v_0 (with $c_{v_0} = 0$). The **Prize Collecting Traveling Salesman Problem** consists of finding a cycle in G containing the node v_0 such that the sum of the edge weights of the cycle minus the sum of the benefits of the nodes of the cycle is minimized. We can get rid of the node weights if we substitute the edge weights c_{ij} by $c_{ij} - \frac{1}{2}v_i - \frac{1}{2}v_j$. Now the prize collection TSP amounts to finding a shortest cycle in G containing v_0. More details are given in BALAS (1989) and RAMESH, YOON & KARWAN (1992).

We have seen that a variety of problems can be transformed to symmetric TSPs or are at least related to it. However, each such transformation has to be considered with some care before actually trying to use it for practical problem solving. E. g., the shortest path computations necessary to treat a GTSP as a TSP take time $O(n^3)$ which might not be acceptable in practice. Most transformations require the introduction of a large number M. This can lead to numerical problems or may even prevent finding feasible solutions at all using only heuristics. In particular, for LP-based approaches, the usage of the "big M" cannot be recommended in general. But, in any case, these transformations provide a basic means for using a TSP code to treat related problems.

3.2 Practical Applications of the TSP

Since we are aiming at the development of algorithms and heuristics for practical traveling salesman problem solving we give a survey on some of the possible applications. The list is not complete but covers the most important cases. In addition we have included problems which cannot be transformed to pure TSPs, but which can be attacked using variants of the methods to be described later.

3.2.1 Drilling of Printed Circuit Boards

The drilling problem for printed circuit boards (PCBs) is a standard application of the symmetric traveling salesman problem. To connect a conductor on one layer with a conductor on another layer or to position (in a later stage of the PCB production) the pins of integrated circuits, holes have to be drilled through the board. The holes may be of different diameters. To drill two holes of different diameters consecutively, the head of the machine has to move to a tool box and change the drilling equipment. This

is quite time consuming. Thus it is clear at the outset that one has to choose some diameter, drill all holes of the same diameter, change the drill, drill the holes of the next diameter etc.
Thus, this drilling problem can be viewed as a sequence of symmetric traveling salesman problems, one for each diameter resp. drill, where the "cities" are the initial position and the set of all holes that can be drilled with one and the same drill. The "distance" between two cities is the time it takes to move the head from one position to the other. The goal is to minimize the travel time for the head of the machine.
We will discuss an application of the drilling problem in depth in Chapter 11.

3.2.2 X-Ray Crystallography

A further direct application of the TSP occurs in the analysis of the structure of crystals (BLAND & SHALLCROSS (1989), DREISSIG & UEBACH (1990)). Here an X-ray diffractometer is used to obtain information about the structure of crystalline material. To this end a detector measures the intensity of X-ray reflections of the crystal from various positions. Whereas the measurement itself can be accomplished quite fast there is a considerable overhead in positioning time since up to 30,000 positions have to be realized for some experiments. In the two examples that we refer to, the positioning involves moving four motors. The time needed to move from one position to the other can be computed very accurately. For the experiment the sequence in which the measurements at the various positions are taken is irrelevant. Therefore, in order to minimize the total positioning time the best sequence for the measurements has to be determined. This problem can be modeled as a symmetric TSP.

3.2.3 Overhauling Gas Turbine Engines

This application was reported by PLANTE, LOWE & CHANDRASEKARAN (1987) and occurs when gas turbine engines of aircrafts have to be overhauled. To guarantee a uniform gas flow through the turbines there are so-called nozzle-guide vane assemblies located at each turbine stage. Such an assembly basically consists of a number of nozzle guide vanes affixed about its circumference. All these vanes have individual characteristics and the correct placement of the vanes can result in substantial benefits (reducing vibration, increasing uniformity of flow, reducing fuel consumption). The problem of placing the vanes in the best possible way can be modeled as a symmetric TSP.

3.2.4 The Order-Picking Problem in Warehouses

This problem is associated with material handling in a warehouse (RATLIFF & ROSENTHAL (1981)). Assume that at a warehouse an order arrives for a certain subset of the items stored in the warehouse. Some vehicle has to collect all items of this order to ship them to the customer. The relation to the TSP is immediately seen. The storage locations of the items correspond to the nodes of the graph. The distance between two nodes is given by the time needed to move the vehicle from one location to the other.

The problem of finding a shortest route for the vehicle with minimal pickup time can now be solved as a TSP. In special cases this problem can be solved easily (see VAN DAL (1992) for an extensive discussion).

3.2.5 Computer Wiring

A special case of connecting components on a computer board is reported in LENSTRA & RINNOOY KAN (1974). Modules are located on a computer board and a given subset of pins has to be connected. In contrast to the usual case where a Steiner tree connection is desired, here the requirement is that no more than two wires are attached to each pin. Hence we have the problem of finding shortest Hamiltonian paths with unspecified starting and terminating points.

A similar situation occurs for the so-called testbus wiring. To test the manufactured board one has to realize a connection which enters the board at some specified point, runs through all the modules, and terminates at some specified point. For each module we also have a specified entering and leaving point for this test wiring. This problem also amounts to solving a Hamiltonian path problem with the difference that the distances are not symmetric and that starting and terminating point are specified.

3.2.6 Clustering of a Data Array

This application is also reported in LENSTRA & RINNOOY KAN (1974). An (r,s)-matrix $A = (a_{ij})$ is given representing relationships between two finite sets of elements $\mathcal{R} = \{R_1, R_2, \ldots, R_r\}$ and $\mathcal{S} = \{S_1, S_2, \ldots, S_s\}$. The entry a_{ij} gives the strength of the relationship between $R_i \in \mathcal{R}$ and $S_j \in \mathcal{S}$. The task is to identify clusters of highly related elements.

To this end, a permutation of the rows and columns of A has to be found which maximizes the sum of all products of horizontally or vertically adjacent pairs of entries of A. We transform this problem as follows.

If ρ and σ are permutations of $\mathcal{R}$ and $\mathcal{S}$, respectively, then the corresponding *measure of effectiveness* is

$$\mathrm{ME}(\rho,\sigma) = \sum_{j=1}^{s-1}\sum_{i=1}^{r} a_{i,\sigma(j)}a_{i,\sigma(j+1)} + \sum_{i=1}^{r-1}\sum_{j=1}^{s} a_{\rho(i),j}a_{\rho(i+1),j}.$$

The two terms can be evaluated separately. We only consider the first one. Let $V = \{1, 2, \ldots, s\}$ and $E = V \times V$. We define weights for the edges of $G = (V, E)$ by

$$c_{ij} = -\sum_{k=1}^{r} a_{ki}a_{kj}.$$

Then the problem of maximizing the first term amounts to finding a shortest Hamiltonian path in G with arbitrary starting and terminating node. The clustering problem is then solved as two separate such problems, one for the rows and one for the columns. The special cases where A is symmetric or where A is a square matrix and where we only allow simultaneous permutations of rows and columns lead to a single Hamiltonian path problem.

3.2.7 Seriation in Archeology

Suppose archeologists have discovered a graveyard and would like to determine the chronological sequence of the various gravesites. To this end each gravesite is classified according to the types of items contained in it. A distance measure between two gravesites is introduced reflecting the diversity between their respective contents. A very likely chronological sequence can be found by computing the shortest Hamiltonian path in the graph whose nodes correspond to the gravesites and where distances are given due to the criterion above. In fact, this was one of the earliest applications mentioned for the TSP.

3.2.8 Vehicle Routing

Suppose that in a city n mail boxes have to be emptied every day within a certain period of time, say 1 hour. The problem is to find the minimal number of trucks to do this and the shortest time to do the collections using this number of trucks. As another example, suppose that customers require certain amounts of some commodities and a supplier has to satisfy all demands with a fleet of trucks. Here we have the additional problem to assign customers to trucks and to find a delivery schedule for each truck so that its capacity is not exceeded and the total travel cost is minimized.
The vehicle routing problem is solvable as a TSP if there is no time constraint or if the number of trucks is fixed (say m). In this case we obtain an m-salesmen problem. Nevertheless, one can apply methods for the TSP to find good feasible solutions for this problem (see LENSTRA & RINNOOY KAN (1974)).

3.2.9 Scheduling

We are given n jobs that have to be performed on some machine. The time to process job j is t_{ij} if i is the job performed immediately before j (if j is the first job then its processing time is t_{0j}). The task is to find an execution sequence for the jobs such that the total processing time is as short as possible.
We define the directed graph $D = (V, A)$ with node set $V = \{0, 1, 2, \ldots, n\}$ and arc set $A = \{1, 2, \ldots, n\} \times \{1, 2, \ldots, n\} \cup \{(0, i) \mid i = 1, 2, \ldots, n\}$. Arc weights are t_{ij} for $(i, j) \in A$. The scheduling problem can now be solved by finding a shortest (directed) Hamiltonian path starting at node 0 with arbitrary terminating node.
Sometimes it is requested that the machine returns to its initial state after having performed the last job. In this case we add arcs $(i, 0)$ for every $i = 1, 2 \ldots, n$ where t_{i0} is the time needed to return to the initial state if job i was performed last. Now the scheduling problem amounts to solving the asymmetric TSP in the new digraph.
Suppose the machine in question is an assembly line and that the jobs correspond to operations which have to be performed on some product at the workstations of the line. In such a case the primary interest would lie in balancing the line. Therefore instead of the shortest possible time to perform all operations on a product the longest individual processing time needed on a workstation is important. To model this requirement a bottleneck TSP is more appropriate.

In LENSTRA & RINNOOY KAN (1974) it is shown that the following job-shop scheduling problem can be transformed to an asymmetric TSP. We are given n jobs that have to be processed on m machines. Each job consists of a sequence of operations (possibly more than m) where each operation has to be performed on one of the machines. The operations have to be performed one after the other in a sequence which is given in advance. As a restriction we have that no passing is allowed (we have the same processing order of jobs on every machine) and that each job visits each machine at least once. The problem is to find a schedule for the jobs that minimizes the total processing time.

3.2.10 Mask Plotting in PCB Production

For the production of each layer of a printed circuit board, as well as for layers of integrated semiconductor devices, a photographic mask has to be produced. In our case for printed circuit boards this is done by a mechanical plotting device. The plotter moves a lens over a photosensitive coated glass plate. The shutter may be opened or closed to expose specific parts of the plate. There are different apertures available to be able to generate different structures on the board. Two types of structures have to be considered. A line is exposed on the plate by moving the closed shutter to one endpoint of the line, then opening the shutter and moving it to the other endpoint of the line. Then the shutter is closed. A point type structure is generated by moving the appropriate aperture to the position of that point then opening the shutter just to make a short flash, and then closing it again. Exact modeling of the plotter control problem leads to a problem more complicated than the TSP and also more complicated than the rural postman problem.
We will discuss an application of the plotting problem in Chapter 11.

3.2.11 Control of Robots

In order to manufacture some workpiece a robot has to perform a sequence of operations on it (drilling of holes of different diameters, cutting of slots, planishing, etc.). The task is to determine a sequence to perform the necessary operations that leads to the shortest overall processing time. A difficulty in this application arises because there are precedence constraints that have to be observed. So here we have the problem of finding the shortest Hamiltonian path (where distances correspond to times needed for positioning and possible tool changes) that satisfies certain precedence relations between the operations. This problem cannot be formulated as a TSP in a straightforward way, but can be treated by applying methods similar to those presented in the forthcoming chapters.

3.3 The Test Problem Instances

Throughout this tract we will use a set of sample problems compiled from various sources to compare the different approaches and to examine the behaviour of algorithms with respect to problem sizes.
To choose a suitable test set one has to decide between contradicting goals.

- Too many computational results might bore the reader.
- Too few results may not exhibit too much insight.
- Problems from different sources might not be comparable in a fair way because distance computations may be more complicated and time consuming in one case.
- Not all problems are suitable for every approach.

To overcome these difficulties we have chosen to proceed as follows.

- We have selected a set of sample problem instances which can always be treated by our methods and which are of the same type. This set consists of twenty-four Euclidean problems in the plane of sizes from 198 to 5934 nodes and is given in Table 3.1. This table also gives the currently best known upper and lower bounds for the respective problems. A number in boldface indicates that an optimal solution is known (and proved!).

Problem	Size	Bounds
d198	198	**15780**
lin318	318	**42090**
fl417	417	**11841**
pcb442	442	**50778**
u574	574	**36905**
p654	654	**34643**
rat783	783	**8806**
pr1002	1002	**259045**
u1060	1060	**224094**
pcb1173	1173	**56892**
d1291	1291	**50801**
rl1323	1323	**270199**
fl1400	1400	[19849,20127]
u1432	1432	**152970**
fl1577	1577	[22137,22249]
d1655	1655	**62128**
vm1748	1748	**336556**
rl1889	1889	**316536**
u2152	2152	[64163,64294]
pr2392	2392	**378032**
pcb3038	3038	**137694**
fl3795	3795	[28594,28772]
fnl4461	4461	**182566**
rl5934	5934	[554070,556146]

Table 3.1 Bounds for sample problems

- In addition we sometimes report about results on other problems. Most of these problem instances are contained in the library TSPLIB of traveling salesman problem instances (see REINELT (1991a)) and are therefore at the disposal of the reader to conduct own experiments.

In our experiments, we have almost completely dispensed of random problem instances. Only at some points we have included a few random problems for extrapolating CPU times. For these problem instances, the points are located in a square and are drawn independently from a uniform distribution. To our opinion, this should be the primary reason for considering random problems. With respect to the assessment of algorithms one should prefer the treatment of real instances (if available) because these are the problems whose solution is of interest. Moreover, real problems have properties that cannot be modeled by random distributions in an appropriate way.

Concerning the CPU times that are either explicitly given or presented in a graphical display the following remarks apply.

- All CPU times are given in seconds on the SUN SPARCstation 10/20. All software (unless otherwise noted) has been written in C and was compiled using `cc` with option `-O4` under the operating system `SUNOS 4.1.2`. The function `get_rusage` was used for measuring running times.

- Distance matrices are not used. Except for precomputed distances for candidate sets, distances are always computed by evaluating the Euclidean distance function and not by a matrix-lookup.

- If a figure displays CPU times for problems up to size 6,000 then these are times for the 24 standard problem instances listed in Table 3.1.

- If a figure displays CPU times for problems up to size 20,000 then this includes in addition the times for further problem instances, in particular for the real problems `rl11849`, `brd14051`, and `d18512`, and for random problems of size 8,000 to 20,000.

Usually, we will not give the explicit length of tours produced by the various heuristics. Rather we give their quality with respect to the lower bounds of Table 3.1. More precisely, if c_{H} is the length of a tour computed by heuristic H and if c_{L} is the lower bound of Table 3.1 for the respective problem instance, we say that the heuristic tour has **quality** $100 \cdot (c_{\mathrm{H}}/c_{\mathrm{L}} - 1)$ percent.

Chapter 4

Geometric Concepts

Many traveling salesman problem instances arising in practice have the property that they are defined by point sets in the 2-dimensional plane (e.g., drilling and plotting problems to be discussed in Chapter 11). Though true distances can usually be not given exactly or only by very complicated formulae, they can very well be approximated by metric distances. To speed up computations we can therefore make use of geometric properties of the point set. In this chapter we introduce concepts for deriving informations about the structure of point sets in the plane and review some basic algorithms. A textbook on computational geometry is EDELSBRUNNER (1987).

4.1 Voronoi Diagrams

Although known since quite some time (VORONOI (1908)) the Voronoi diagram has only recently received the attention of many researchers in the field of computational geometry. It has several attractive features and puts particular emphasis on proximity relations between points.

Let $S = \{P_1, P_2, \ldots, P_n\}$ be a finite subset of $\mathbf{R}^m$ and let $d : \mathbf{R}^m \times \mathbf{R}^m \longrightarrow \mathbf{R}$ be a metric. We define the **Voronoi region** $\mathrm{VR}(P_i)$ of a point P_i via

$$\mathrm{VR}(P_i) = \{P \in \mathbf{R}^m \mid d(P, P_i) \le d(P, P_j) \text{ for all } j = 1, 2, \ldots, n,\ j \ne i\},$$

i.e., $\mathrm{VR}(P_i)$ is the set of all points that are at least as close to P_i as to any other point of S. The set of all n Voronoi regions is called the **Voronoi diagram** $\mathrm{VD}(S)$ of S. Other names are **Dirichlet tessellation** or **Thiessen tessellation.** In the following we call the elements of S **generators.**

We consider only the 2-dimensional case. With each generator P_i we associate its Cartesian coordinates (x_i, y_i). For the first part we assume that d is the Euclidean metric (L_2), i.e.,

$$d((x_1, y_1), (x_2, y_2)) = \sqrt{(x_1 - x_2)^2 + (y_1 - y_2)^2}.$$

For two generators P_i and P_j we define the **perpendicular bisector** $B(P_i, P_j) = \{P \in \mathbf{R}^2 \mid d(P, P_i) = d(P, P_j)\}$. If we define the half space $B_{ij} = \{x \in \mathbf{R}^2 \mid d(P_i, x) \le d(P_j, x)\}$ then we see that

$$\mathrm{VR}(P_i) = \bigcap_{\substack{j=1 \\ j \ne i}}^{n} B_{ij}.$$

This implies that in the case of the Euclidean metric the Voronoi regions are convex polygons with at most $n-1$ vertices. Figure 4.1 shows the Voronoi diagram for the points defining the traveling salesman problem instance rd100, a random problem on 100 points.

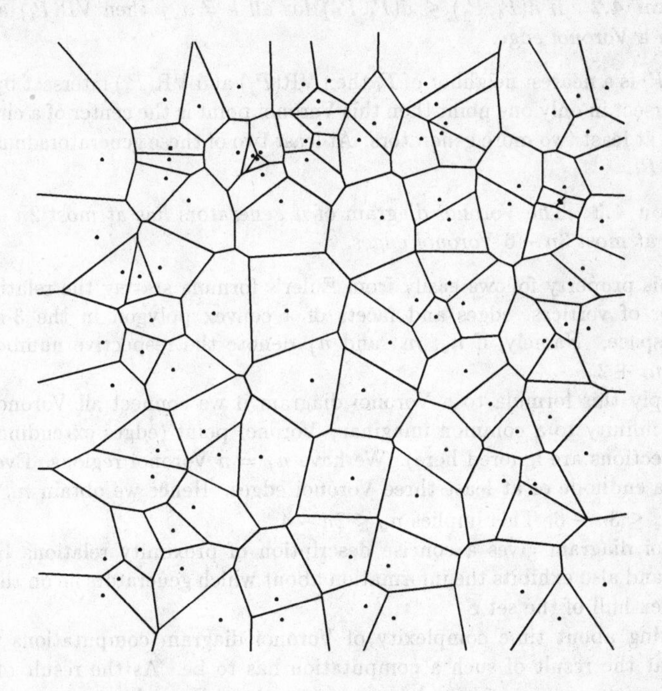

Figure 4.1 Voronoi diagram for rd100 (L_2-metric)

We call nonempty intersections of two or more Voronoi regions **Voronoi points** if they have cardinality 1 and **Voronoi edges** otherwise. Note that every Voronoi point is the center of a circle through (at least) three generators and every Voronoi edge is a segment of a perpendicular bisector of two generators.

A generator configuration is called **degenerate** if there are four generators lying on a common circle. We say that two Voronoi regions are **adjacent** if they intersect. Note that only in degenerate cases two regions can intersect in just one point. Degeneracy does not occur in Figure 4.1 as is usually the case for randomly generated points. Some observations are important.

Proposition 4.1 *The Voronoi diagram of a generator P_i is unbounded if and only if P_i lies on the boundary of the convex hull of S.*

Proof. If $\mathrm{VR}(P_i)$ is bounded then P_i is contained in the interior of the convex hull of those generators whose regions are adjacent to $\mathrm{VR}(P_i)$. On the other hand, if $\mathrm{VR}(P_i)$ is unbounded it cannot be situated in the interior of $\mathrm{conv}(S)$. □

An unbounded region does not necessarily imply that a generator is a vertex of the convex hull. Consider for example the case where all generators are located on a line. Then all regions are unbounded, but only two generators define the convex hull of the set.

Proposition 4.2 *If $d(P_i, P_j) \leq d(P_i, P_k)$ for all $k \neq i, j$ then $VR(P_i)$ and $VR(P_j)$ intersect in a Voronoi edge.*

Proof. If P_j is a nearest neighbor of P_i then $\mathrm{VR}(P_j)$ and $\mathrm{VR}(P_i)$ intersect by definition. If they intersect in only one point then this Voronoi point is the center of a circle through P_i, P_j, and at least two more generators. At least two of these generators must be nearer to P_i than P_j. □

Proposition 4.3 *The Voronoi diagram of n generators has at most $2n - 4$ Voronoi points and at most $3n - 6$ Voronoi edges.*

Proof. This property follows easily from Euler's formula stating the relation between the number of vertices, edges and facets of a convex polygon in the 3-dimensional Euclidean space. Namely, if n_v, n_e, and n_f denote the respective numbers we have $n_f + n_v = n_e + 2$.
We can apply this formula to a Voronoi diagram if we connect all Voronoi edges extending to infinity to a common imaginary Voronoi point (edges extending to infinity in both directions are ignored here). We have $n_f = n$ Voronoi regions. Every Voronoi point is the endnode of at least three Voronoi edges. Hence we obtain $n_v \leq \frac{2}{3} n_e$ and therefore $n_e \leq 3n - 6$. This implies $n_v \leq 2n - 4$. □
The Voronoi diagram gives a concise description of proximity relations between the generators and also exhibits the information about which generators lie on the boundary of the convex hull of the set S.
Before talking about time complexity of Voronoi diagram computations we have to specify what the result of such a computation has to be. As the result of a Voronoi diagram algorithm we require a data structure that allows for an easy access to the Voronoi edges forming a specific region as well as to all Voronoi edges containing a given Voronoi point. This can, e.g., be achieved by a data structure proposed in OTTMANN & WIDMAYER (1990). We store a doubly linked list of the Voronoi edges together with the following information for every edge:

- tail and head nodes, u and v, of the edge (the number -1 indicates that the edge extends to infinity),
- the Voronoi regions V_l and V_r to the left and to the right of the edge (where left and right are with respect to some arbitrarily imposed direction of the edge),
- pointers to the next Voronoi edge of region V_l (resp. V_r) having u (resp. v) as one endnode.

A straightforward algorithm for computing the Voronoi diagram takes time $O(n^2)$. We just compute each $\mathrm{VR}(P_i)$ as the intersection of the halfspaces B_{ij} for $j \neq i$.
A lower bound on the running time of every Voronoi diagram algorithm is established as follows. Consider the situation that all generators are located on the x-axis, i.e., their y-coordinates are 0. If we have the Voronoi diagram of this set we easily obtain the sorted sequence of the generators with respect to their x-coordinates. Since sorting

cannot be performed in less than $O(n \log n)$ time we have a lower bound on the running time for Voronoi diagram computations. Note that sorting and computing the Voronoi diagram are essentially equivalent in this 1-dimensional case. The following section will describe an algorithm that achieves this best possible running time.
In the present and in the following sections which also deals with geometric problems we will not discuss the algorithms in full detail. When solving geometric problems very often certain "degenerate" situations can occur that may have the consequence that some object or some operation is not uniquely specified. Explaining the idea of a geometric algorithm is often easy but when it comes to implementing the algorithm one will usually face severe difficulties. It is a commonly observed fact that, when coding algorithms for geometric problems, about 10–20% of the code account for the "real" algorithm while the remaining part is only necessary to deal with degenerate situations. Almost the same is true when describing such algorithms in detail. We will therefore not discuss degenerate situations explicitly, but will assume that points defining a geometric problem are in general position, i.e., no degeneracy occurs. The reader should have in mind, however, that an implementation must take care of these cases.
The $O(n \log n)$ algorithm to be described now was given in SHAMOS AND HOEY (1975). It is a divide and conquer approach that divides the set of generators recursively into two halves, computes the respective Voronoi diagram for the two parts, and merges them to obtain the Voronoi diagram of the whole set S.

procedure divide_and_conquer

(1) If the current set has at most three elements compute its Voronoi diagram and return.

(2) Partition S into two sets $S_1 = \{P_{i_1}, P_{i_2}, \ldots P_{i_l}\}$ and $S_2 = \{P_{i_{l+1}}, P_{i_{l+2}}, \ldots, P_{i_n}\}$ where $l = \lfloor \frac{n}{2} \rfloor$ such that there is a vertical line separating S_1 and S_2.

(3) Perform the divide and conquer algorithm recursively to compute $\mathrm{VD}(S_1)$ and $\mathrm{VD}(S_2)$.

(4) Obtain $\mathrm{VD}(S)$ from $\mathrm{VD}(S_1)$ and $\mathrm{VD}(S_2)$.

end of divide_and_conquer

The partition in Step (2) can be found in linear time if we have sorted the elements of S in a preprocessing step with respect to their x-coordinates. There is also a more complicated algorithm achieving linear running time without preprocessing (see Chapter 2 for median finding).
The critical step is Step (4). If this step can be performed in linear time (linear in $|S_1| + |S_2|$) then the basic recurrence relation for divide and conquer algorithms gives a time bound of $O(n \log n)$ for the overall Voronoi computation.
Merging two Voronoi diagrams (where one set of generators is to the left of the other set of generators) basically consists of identifying the thick "merge line" in the center of Figure 4.2.
We cannot go into detail here and only discuss the principle of the construction of this merge line. The line is constructed from the bottom to the top. One first has to identify the respective generators in $\mathrm{VD}(S_1)$ and $\mathrm{VD}(S_2)$ with minimal y-coordinates

(and therefore unbounded Voronoi regions). The perpendicular bisector between these two generators then gives the lower part of the merge line. The process of merging the two diagrams can now be visualized as extending the merge line upwards until an edge of either of the two Voronoi diagrams is hit. At such a point the merge line may change direction because it will now be part of the perpendicular bisector between two other generators. This process is continued until the final part of the merge line is part of the bisector of the two generators in the respective diagrams with maximal y-coordinates.

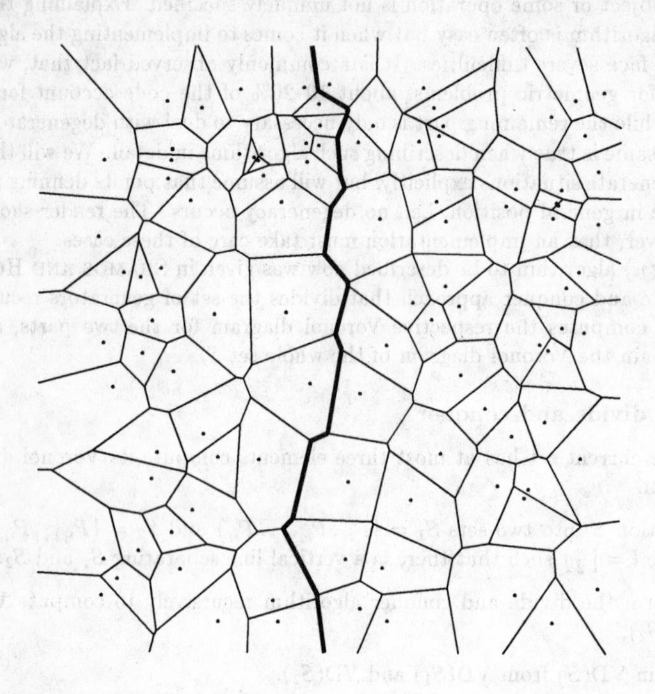

Figure 4.2 The merge line for merging two Voronoi diagrams

Implementing these steps carefully results in a running time of $O(|S_1| + |S_2|)$ for the merge process. This gives the desired optimal running time of the divide and conquer algorithm.
The principle of the next algorithm (see GREEN & SIBSON (1978)) is to start out with the Voronoi diagram for three generators and then to successively take into account the remaining generators and update the current Voronoi diagram accordingly.

procedure incremental_algorithm

(1) Compute the Voronoi diagram $\mathrm{VD}(\{P_1, P_2, P_3\})$.

(2) For $t = 4, 5, \ldots, n$ compute the diagram $\mathrm{VD}(\{P_1, P_2, \ldots, P_t\})$ as follows.

(2.1) Find P_s, $1 \leq s \leq t-1$, such that $P_t \in \mathrm{VR}(P_s)$, i.e., find the nearest neighbor of P_t among the generators already considered.

(2.2) Start with the perpendicular bisector of P_t and P_s and find its intersection with a boundary edge of $VR(P_s)$. Suppose this edge is also the boundary edge of $VR(P_i)$. Find the other intersection of the bisector between P_t and P_i with the boundary of $VR(P_i)$. Proceed this way by entering neighboring regions and computing intersections between the current bisector $B(P_t, P_j)$ and the boundary of $VR(P_j)$ until the starting region $VR(P_s)$ is reached again. Eliminate everything within the resulting closed walk. (In the case of an unbounded region $VR(P_t)$ some further details have to be observed.)

end of incremental_algorithm

Figure 4.3 shows a typical step of the incremental algorithm. The broken lines indicate the bisectors that are followed in Step (2.2). The new diagram is obtained by deleting everything inside the convex polygon determined by the broken lines.

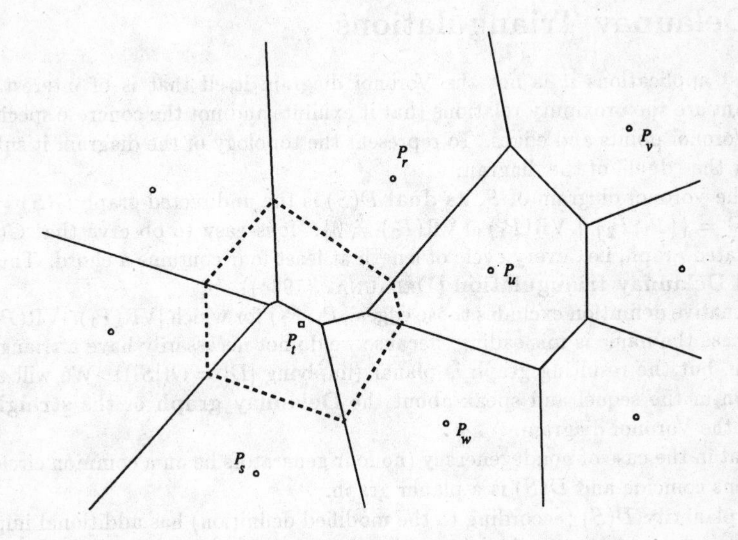

Figure 4.3 A step of the incremental algorithm

The running time of this algorithm can only be bounded above by $O(n^2)$. In OHYA, IRI & MUROTA (1984) the algorithm is examined for application in practice. It turns out that an enormous speed up can be obtained if the generators are considered in a clever sequence. Ohya et al. use a bucketing approach which allows the nearest neighbor in Step (2.1) to be guessed with high probability in constant time. Step (2.2) is not that critical because usually only very local updates have to be performed to obtain the new Voronoi diagram. In practical experiments linear running time on several classes of randomly generated point sets was observed. In particular, the incremental algorithm outperformed the divide and conquer method.

It is interesting to note, that selecting the next generator at random results in observed $O(n^{\frac{3}{2}})$ running time.

Expected linear running time is not proven in OHYA, IRI & MUROTA (1984), but some evidence is given for it. For generators that are independently drawn from a uniform distribution on the unit square, the expected number of generators to be examined in Step (2.1) to find the nearest neighbor is bounded by a constant. In addition, the expected number of Voronoi edges of a Voronoi region in any intermediate diagram is bounded by a constant. These are two basic results that suggest that also a rigorous mathematical proof of linear expected running time for this algorithm can be obtained.
A further algorithm for Voronoi diagram construction has been given in FORTUNE (1987). This algorithm uses a sweep-line principle to compute the diagram in time $O(n \log n)$.
CRONIN (1990) and RUJÁN, EVERTSZ AND LYKLEMA (1988) employ the Voronoi diagram for generating traveling salesman tours.

4.2 Delaunay Triangulations

For most applications it is not the Voronoi diagram itself that is of interest. More important are the proximity relations that it exhibits and not the concrete specification of the Voronoi points and edges. To represent the topology of the diagram it suffices to consider the "dual" of the diagram.
Given the Voronoi diagram of S, its **dual** $D(S)$ is the undirected graph $G(S) = (S, D)$ where $D = \{\{P_1, P_2\} \mid \text{VR}(P_1) \cap \text{VR}(P_2) \neq \emptyset\}$. It is easy to observe that $G(S)$ is a triangulated graph, i.e., every cycle of length at least four contains a chord. This graph is called **Delaunay triangulation** (DELAUNAY (1934)).
An alternative definition excludes those edges $\{P_1, P_2\}$ for which $|\text{VR}(P_1) \cap \text{VR}(P_2)| = 1$. In this case the name is misleading, because we do not necessarily have a triangulation anymore, but the resulting graph is planar (implying $|D| = O(|S|)$). We will use this definition in the sequel and speak about the **Delaunay graph** or the **straight line dual** of the Voronoi diagram.
Note that in the case of nondegeneracy (no four generators lie on a common circle) both definitions coincide and $D(S)$ is a planar graph.
Besides planarity $D(S)$ (according to the modified definition) has additional important properties.

Proposition 4.4

(i) If P_i and P_j are generators such that $d(P_i, P_j) \leq d(P_i, P_k)$ for all $k \neq i, j$ then $\{P_i, P_j\}$ is an edge of $D(S)$.

(ii) $D(S)$ has at most $3n - 6$ edges.

(iii) $D(S)$ contains a minimum spanning tree of the complete graph on n nodes where the nodes correspond to the generators and the edge weights are respective Euclidean distances.

Proof. Part (i) is clear because of Proposition 4.2 and part (ii) follows from Proposition 4.3.
For part (iii) consider Prim's algorithm to compute a minimum spanning tree. In each step we have a set V of nodes that are already connected by a spanning tree and the

set $S \setminus V$ (consisting of isolated nodes). The next edge to be added to the tree is the shortest edge connecting a node in V to a node in $S \setminus V$. This edge must be contained in $D(S)$ since it connects two generators whose Voronoi regions intersect in a Voronoi edge. □

Figure 4.4 shows the Delaunay triangulation corresponding to the diagram of Figure 4.1.

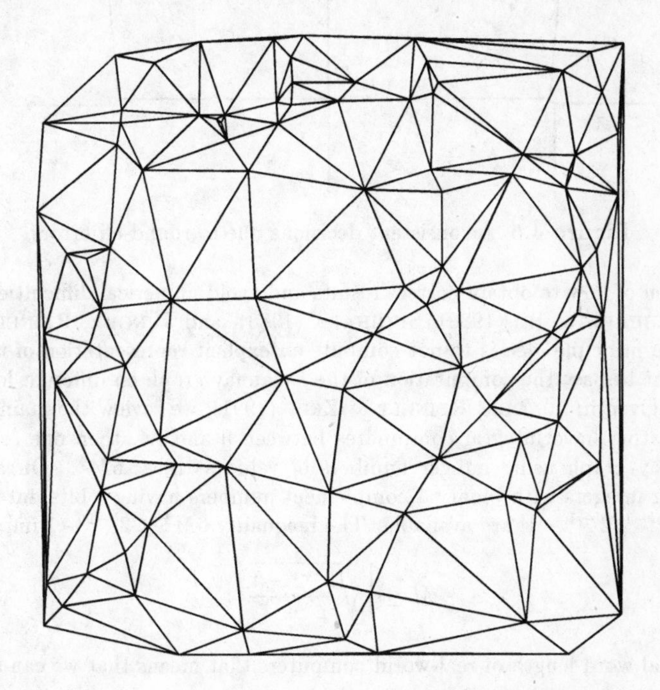

Figure 4.4 The Delaunay triangulation for rd100

Note that Proposition 4.4 (ii) does not hold for the general Delaunay triangulation but only for the Delaunay graph. For example, if all generators are located on a circle then all Voronoi regions intersect in a common Voronoi point and the Delaunay triangulation is the complete graph on n nodes. The Delaunay graph is only a cycle of length n.

Straightforward implementations of algorithms for computing the Voronoi diagram (or the Delaunay triangulation), in which all numerical computations are carried out in floating point arithmetic, run into numerical problems.

Voronoi points are given as intersection points of bisectors. Due to insufficient accuracy it may not be possible to safely decide whether two lines are parallel or almost parallel. Moreover, the intersection points may be located "far away" from the generators leading to imprecise computations of Voronoi points.

The consequence is that due to incorrect decisions the algorithm may not work because computed data is contradictory.

Consider the following example (Figure 4.5). We have three generators that are located at three corners of a square. Depending on whether the fourth generator is at the fourth corner, or inside, or outside the square different Voronoi diagrams arise. If it cannot be exactly differentiated between these three cases, then the correct computation of the Voronoi diagram and hence of the Delaunay triangulation fails.

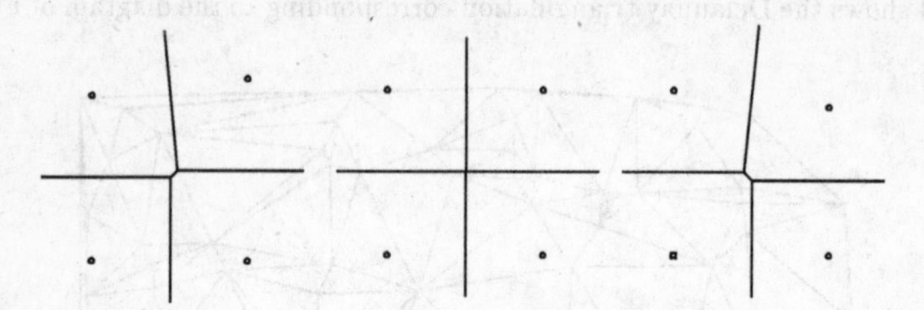

Figure 4.5 Inconsistent decisions due to round-off errors

The question of how to obtain correct results and avoid numerical difficulties is considered in SUGIHARA & IRI (1988), SUGIHARA (1988), and JÜNGER, REINELT & ZEPF (1991). The principle idea is to not compute an explicit representation of the Voronoi diagram, but to base the computation of the Delaunay graph on different logical tests. Details are given in JÜNGER, REINELT & ZEPF (1991), we review the main results.
If all generators have integral coordinates between 0 and M, then one can compute the Delaunay graph using integer numbers of value at most $6M^4$. On a computer representing integers with binary 1-complement numbers having b bits, integers in the interval $[-2^{b-1}, 2^{b-1} - 1]$ are available. The inequality $6M^4 \leq 2^{b-1} - 1$ implies

$$M \leq \left\lfloor \sqrt[4]{\frac{2^{b-1} - 1}{6}} \right\rfloor.$$

For the usual word length of real-world computers that means that we can allow

$$M \leq \begin{cases} 8 & \text{if } b = 16 \\ 137 & \text{if } b = 32 \\ 35,211 & \text{if } b = 64\,. \end{cases}$$

So, only 32-bit integer arithmetic is not enough for computing correct Delaunay triangulations in practical applications. Only by using at least 64-bit arithmetic we can treat reasonable inputs.
Of special interest are also the two metrics

- **Manhattan metric** (L_1): $d((x_1, y_1), (x_2, y_2)) = |x_1 - x_2| + |y_1 - y_2|$, and
- **Maximum metric** (L_∞): $d((x_1, y_1), (x_2, y_2)) = \max\{|x_1 - x_2|, |y_1 - y_2|\}$.

This is due to the fact that very often distances correspond to the time a mechanical device needs to travel from one point to the other. If this movement is performed first in the horizontal direction and then in the vertical direction the L_1-metric should be chosen to approximate travel times. If the movement is performed by two motors working simultaneously in horizontal and vertical directions then the L_∞-metric is the appropriate choice for modeling the movement times.

For these metrics, bisectors are no longer necessarily straight lines. They may consist of three line segments and we can also have degenerate situations as shown in Figure 4.6.

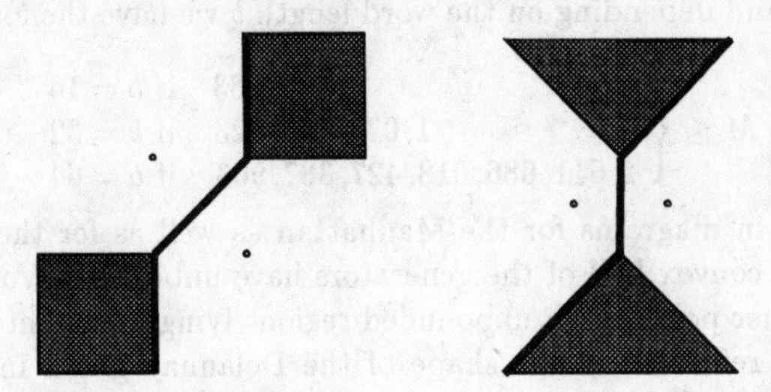

Figure 4.6 Bisectors for L_1- and L_∞-metric

Here the bisectors include the shaded regions. In the L_1-metric (left) this situation occurs when the coordinate differences in both coordinates are the same. In the L_∞-metric (right picture) we have this situation if both points coincide in one of the two coordinates. It is convenient to restrict our definition of bisectors to the bold line segments in Figure 4.6 (the definition of Voronoi regions is changed accordingly). The L_1-metric Voronoi diagram for problem rd100 is shown in Figure 4.7.

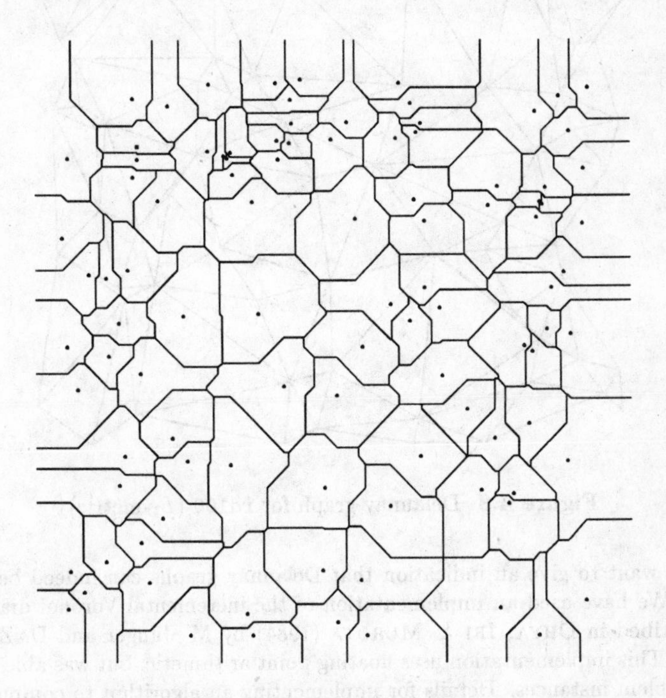

Figure 4.7 Voronoi diagram for rd100 (L_1-metric)

The numerical analysis of the L_1-case shows that we can compute an explicit representation of the Voronoi diagram itself using only one additional bit of accuracy than needed to input the data. It follows that we can carry out all computations with numbers of size at most $2M$, and depending on the word length b we have the following constraints for M

$$M \leq \begin{cases} 16,383 & \text{if } b = 16 \\ 1,073,741,823 & \text{if } b = 32 \\ 4,611,686,018,427,387,903 & \text{if } b = 64\,. \end{cases}$$

Observe that also in diagrams for the Manhattan as well as for the maximum metric the vertices of the convex hull of the generators have unbounded Voronoi regions. But there may be further points with unbounded regions lying in the interior of the convex hull. This is also reflected by the shape of the Delaunay graph in Figure 4.8 which corresponds to the the Voronoi diagram of Figure 4.7.

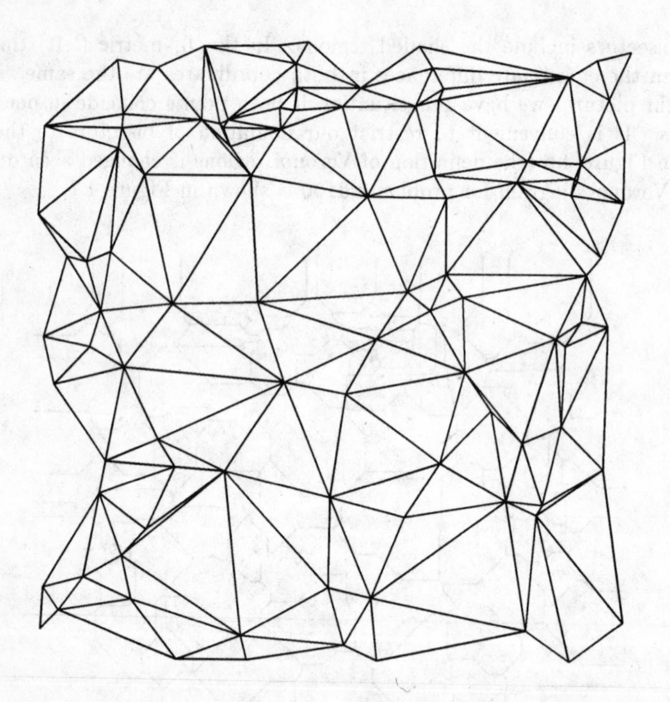

Figure 4.8 Delaunay graph for rd100 (L_1-metric)

Finally, we want to give an indication that Delaunay graphs can indeed be computed very fast. We have used an implementation of the incremental Voronoi diagram algorithm described in OHYA, IRI & MUROTA (1984) by M. Jünger and D. Zepf for the L_2-metric. This implementation uses floating point arithmetic, but was able to solve all sample problem instances. Details for implementing an algorithm to compute Voronoi diagrams for further metrics are discussed in KAIBEL (1993).

Figure 4.9 shows the running time of this implementation on our set of sample problems. CPU times are given in seconds on a SUN SPARCstation 10/20.

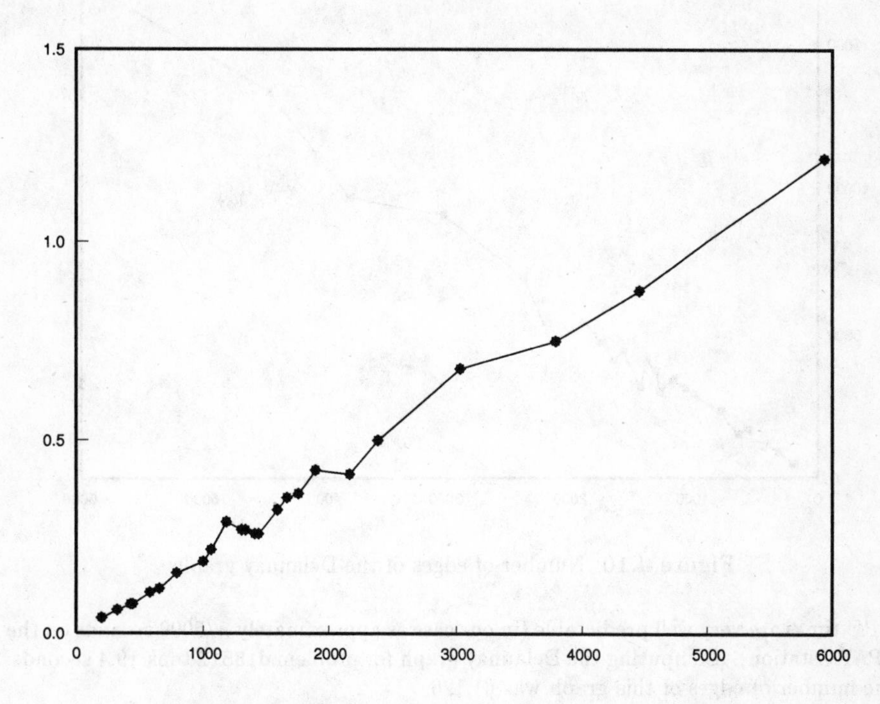

Figure 4.9 CPU times for computing Delaunay graphs

The figure seems to suggest indeed a linear increase of the running times with the problem size. But, as we shall see throughout this monograph, we have to accept that real-world problems do not behave well in the sense that smooth running time functions can be obtained. Just the number of nodes of a problem is not sufficient to characterize it. Real problems have certain structural properties that cannot be modeled by random problem instances and can lead to quite different running times of the same algorithm for different problem instances of the same size.
The number of edges of the respective Delaunay graphs is shown in Figure 4.10.
For random problems the expected number of edges forming a Voronoi region is six, which is hence also the expected degree of a node in the Delaunay graph. Therefore we would expect about $3n$ edges in a Delaunay graph which is quite closely achieved for our sample problems.
We can conclude that Delaunay graphs can be computed very efficiently. For practical purposes it is also important that running times are quite stable even though we have problem instances from various sources and with different structural properties.

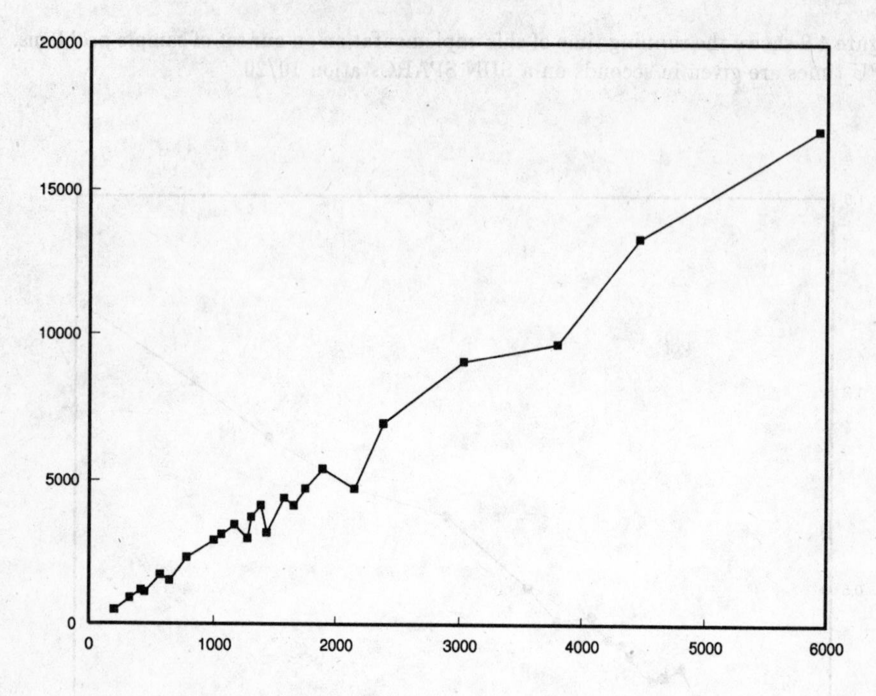

Figure 4.10 Number of edges of the Delaunay graphs

CPU times are very well predictable (in our case as approximately $n/5000$ seconds on the SPARCstation). Computing the Delaunay graph for problem d18512 took 19.4 seconds, the number of edges of this graph was 61,126.

4.3 Convex Hulls

The convex hull of a set of points is the smallest convex set containing these points. It is a convenient means for representing point sets. If the point set is dense then the convex hull may very well reflect its shape. Large instances of traveling salesman problems in the plane usually exhibit several clusters. Building the convex hull of these clusters can result in a concise representation of the whole point set still exhibiting many of its geometric properties.

A short description of complicated objects is also important in other areas, for example in computer graphics or control of robots. Here movements of objects in space have to be traced in order to avoid collisions. In such cases convex hulls can be applied to represent the objects approximately.

We define the problem to compute the **convex hull** as follows. We are given a finite set $A = \{a_1, a_2, \ldots, a_n\}$ of n points in the plane where $a_i = (x_i, y_i)$. The task is to identify those points that constitute the vertices of the convex hull conv(A) of the n points, i.e., which are not representable as a convex combination of other points. Moreover, we require that the computation also delivers the sequence of these vertices in correct order. This means that if the computation outputs the vertices $v_1, v_2, \ldots, v_t$ then the convex hull is the (convex) polygon that is obtained by drawing edges from v_1 to v_2, from v_2 to v_3, etc., and finally by drawing the edges from v_{t-1} to v_t and from v_t to v_1. We will review some algorithms for computing the convex hull that also visualize some fundamental principles in the design of geometric algorithms.

Before starting to describe algorithms we want to make some considerations concerning the time that is at least necessary to find convex hulls. Of course, since all n input points have to be taken into account we must spend at least time of linear order in n. It is possible to establish a better lower bound. Let $B = \{b_1, b_2, \ldots, b_n\}$ be a set of distinct positive real numbers. Consider the set $A \subseteq \mathbf{R}^2$ defined by $A = \{(b_i, b_i^2) \mid 1 \leq i \leq n\}$. Since the function $f : \mathbf{R} \to \mathbf{R}$ with $f(x) = x^2$ is strictly convex none of the points of A is a convex combination of other points. Hence computing the convex hull of A also sorts the numbers b_i. It is known that in many computational models sorting of n numbers needs worst case time $\Omega(n \log n)$.

A second way to derive a lower bound is based on the notion of maxima of vectors. Let $A = \{a_1, a_2, \ldots, a_n\}$ be a finite subset of $\mathbf{R}^2$. We define a partial ordering "$\succ$" on A by

$$a_i = (x_i, y_i) \succ a_j = (x_j, y_j) \text{ if and only if } x_i \geq x_j \text{ and } y_i \geq y_j.$$

The maxima with respect to this ordering are called **maximal vectors**. In KUNG, LUCCIO & PREPARATA (1975) the worst case time lower bound of $\Omega(n \log n)$ for identifying the maximal vectors of A is proved. This observation is exploited in PREPARATA & HONG (1977) for the convex hull problem. Suppose A is such that every a_i is a vertex of the convex hull. Identify the (w.l.o.g.) four points with maximal, resp. minimal x- or y-coordinate. Let a_j be the vertex with maximal y-coordinate and a_k be the vertex with maximal x-coordinate. The number of vertices between a_j and a_k may be of order n. They are all maximal elements. Since convex hull computations can identify maximal vectors it cannot be faster in the worst case than $O(n \log n)$. Note that this lower bound is also valid if we do not require that the vertices of the convex hull are output in their correct sequence.

Though the lower bound derived here may seem to be weak there are many algorithms that compute the convex hull in worst case time $O(n \log n)$.

According to TOUSSAINT (1985) the first efficient convex hull algorithm has been outlined in BASS & SCHUBERT (1967). Their algorithm was designed to be the first step for computing the smallest circle containing a given set of points in the plane. Though the algorithm is not completely correct it already exhibits some of the powerful ideas used in convex hull algorithms. It consists of an elimination step as in the throw-away algorithm of section 4.3.3 and afterwards basically performs a scan similar to Graham's scan (to be described next). When corrected appropriately a worst case running time of $O(n \log n)$ can be shown. Therefore, this algorithm can be considered as the first $O(n \log n)$ convex hull algorithm.

4.3.1 Graham's Scan

This algorithm was given by GRAHAM (1972). Its main step consists of computing a suitable ordering of the points. Then the convex hull is built by successively scanning the points in this order. The algorithm works as follows.

procedure graham_scan

(1) Identify an interior point of conv(A), say P_0. This can be done by finding three points of A that are not collinear and by taking their center of gravity as P_0.

(2) Compute polar coordinates for the points of A with respect to the center P_0 and some arbitrary direction representing the angle zero.

(3) Sort the points with respect to their angles.

(4) If there are points with the same angle then eliminate all of them but the one with largest radius. Let $a_{i_1}, a_{i_2}, \ldots, a_{i_t}$ be the sorted sequence of the remaining points.

(5) Start with three consecutive points P_r, P_m, and P_l, i.e., $P_r = a_{i_k}$, $P_m = a_{i_{k+1}}$, $P_l = a_{i_{k+2}}$ for some index k where indices are taken modulo t.

(6) Perform the following step for the current three points until the same triple of points occurs for the second time.

a) If P_m lies on the same side of the segment $[P_l, P_r]$ as P_0 or lies on the segment then delete P_m and set $P_m = P_r$ and P_r to its predecessor in the current sorted list.

b) If P_m lies on the side of the segment $[P_l, P_r]$ opposite to P_0 then set $P_r = P_m$, $P_m = P_l$, and P_l to its successor in the current sorted list.

end of graham_scan

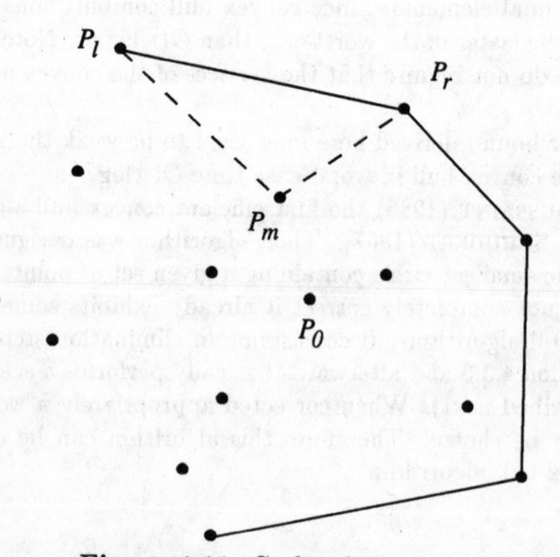

Figure 4.11 Graham's scan

Correctness of this algorithm is easily verified. Step (6) has to be performed at most $2t$ times for scanning the necessary triples of nodes. In Step (6a) a point is discarded so it cannot be performed more than $t-3$ times. If Step (6b) is executed t times we have scanned the points "once around the clock" and no further changes are possible. Therefore the worst case running time is dominated by the sorting in Step (3) and we obtain the worst case running time $O(n \log n)$.
This way we have established the worst case time complexity $\Theta(n \log n)$ for computing convex hulls in the plane.

4.3.2 Divide and Conquer

The divide and conquer principle also applies in the case of convex hull computations (BENTLEY & SHAMOS (1978)). In this case, the basic step consists of partitioning a point set according to some rule into two sets of about equal size, computing their respective convex hulls and merging them to obtain the convex hull of the whole set.

procedure divide_and_conquer

(1) If the current set has at most three elements compute its convex hull and return.

(2) Partition A into two sets $A_1 = \{a_{i_1}, a_{i_2}, \ldots, a_{i_l}\}$ and $A_2 = \{a_{i_{l+1}}, a_{i_{l+2}}, \ldots, a_{i_n}\}$ where $l = \lfloor \frac{n}{2} \rfloor$ such that there is a vertical line separating A_1 and A_2.

(3) Perform the divide and conquer algorithm recursively to compute $\text{conv}(A_1)$ and $\text{conv}(A_2)$.

(4) Merge the two convex hulls to obtain the convex hull of A.

end of divide_and_conquer

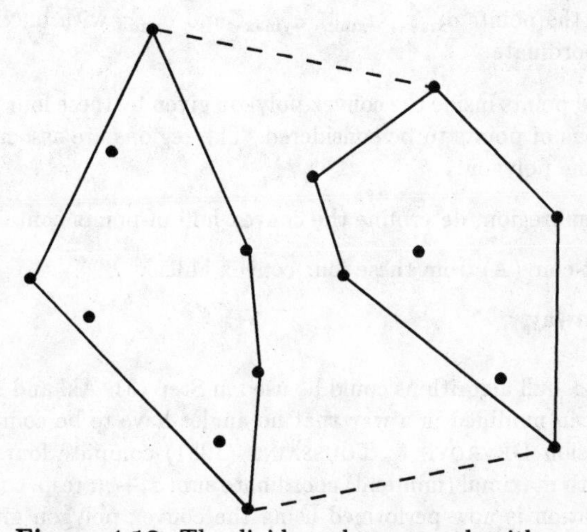

Figure 4.12 The divide and conquer algorithm

The partition required in Step (2) can be easily computed if the points are presorted (which takes time $O(n \log n)$). So, the only critical step to be examined is Step (4). When merging two hulls we have to add two edges (the so-called **upper** and **lower bridges**) and eliminate all edges of conv(A_1) and conv(A_2) that are not edges of conv(A). To find these bridges we can exploit the fact that due to the sorting step we know the leftmost point of A_1 and the rightmost point of A_2 (w.l.o.g., A_1 is left of A_2). Starting at these points it is fairly simple to see that the edges to be eliminated can be readily identified and that no other edges are considered for finding the bridges. Therefore, the overall time needed to merge convex hulls during the algorithm is linear in n. Due to Theorem 2.9, this establishes the $O(n \log n)$ worst case time bound for the divide and conquer approach.
KIRKPATRICK & SEIDEL (1986) describe a refinement of the divide and conquer approach to derive a convex hull algorithm which has worst case running time $O(n \log v)$ where v is the number of vertices of the convex hull.

4.3.3 Throw-Away Principles

It is intuitively clear that when computing the convex hull of a set A not all points are equally important. With high probability, points in the "interior" of A will not contribute to the convex hull whereas points near the "boundary" of A are very likely vertices of the convex hull. Several approaches make use of this observation in that they eliminate points before starting the true convex hull computation.
If we consider a convex polygon whose vertices are contained in the set A then all points inside this polygon can be discarded since they cannot contribute to the convex hull. In AKL & TOUSSAINT (1978) an algorithm is given that makes use of this fact.

procedure throw_away

(1) Compute the points a_{xmax}, a_{xmin}, a_{ymax}, and a_{ymin} with maximal (minimal) x-, resp. y-coordinate.

(2) Discard all points inside the convex polygon given by these four points and identify four regions of points to be considered. The regions are associated with the four edges of the polygon.

(3) For each subregion, determine the convex hull of points contained in it.

(4) Construct conv(A) from these four convex hulls.

end of throw_away

Any of the convex hull algorithms could be used in Step (3). Akl and Toussaint basically use Graham's scan modified in a way that no angles have to be computed.
In a refined version DEVROYE & TOUSSAINT (1981) compute four additional points, namely those with maximal (minimal) coordinate sum x_i+y_i, resp. coordinate difference $x_i - y_i$. Elimination is now performed using the convex polygon given by these eight points.

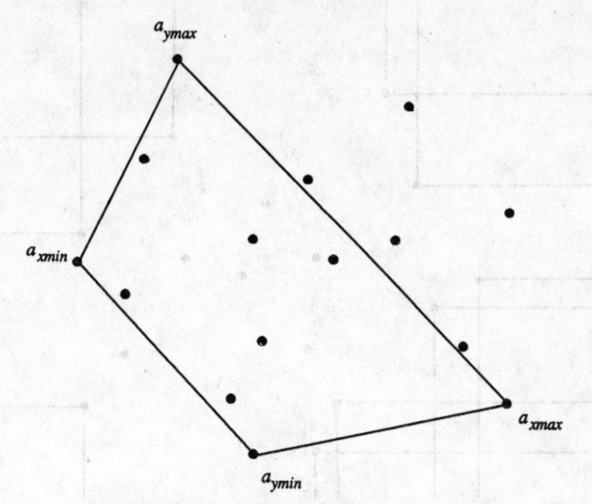

Figure 4.13 A throw-away principle

4.3.4 Convex Hulls from Maximal Vectors

Every point in $\mathbf{R}^2$ can be viewed as the origin of a coordinate system with axes parallel to the x- and y-directions. This coordinate system induces four quadrants. A point is called **maximal** with respect to the set A if at least one of these quadrants (including the axes) does not contain any other point of A. It is easily seen that every vertex of conv(A) is maximal. Namely, let x be a vertex of conv(A) and assume that each of the corresponding quadrants contains a point of A. Then x is contained in the convex hull of these points and therefore cannot be a vertex of the convex hull of A. KUNG, LUCCIO & PREPARATA (1975) give an algorithm to compute the maximal vectors of a point set in the plane in worst case time $O(n \log n)$. This leads to the following $O(n \log n)$ algorithm for computing the convex hull.

procedure maximal_vector_hull

(1) Compute the set S of maximal vectors with respect to A.

(2) Let $A' = S$.

(3) Compute the convex hull of A' using any of the $O(n \log n)$ worst case time algorithms.

(4) $\text{conv}(A) = \text{conv}(A')$.

end of maximal_vector_hull

The expectation is that very many points can be discarded in Step (1).

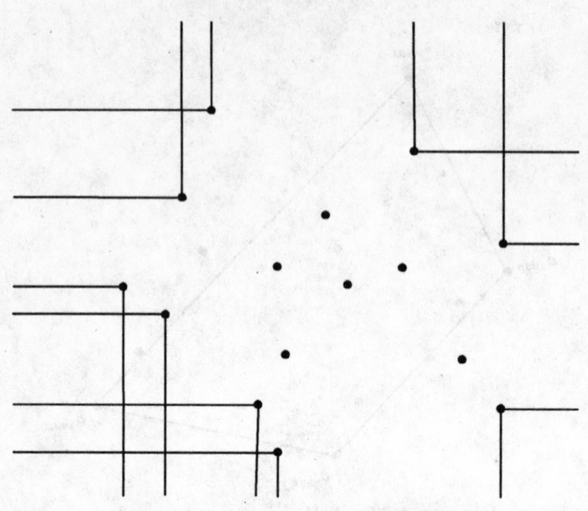

Figure 4.14 Maximal vectors

4.3.5 A New Elimination Type Algorithm

We are now going to discuss a further elimination type algorithm that uses a particularly simple discarding mechanism. This algorithm is best suited for large dense point sets distributed uniformly in a rectangle. It is discussed in full detail in BORGWARDT, GAFFKE, JÜNGER & REINELT (1991).

Assume that the points of A are contained in the unit sqaure, i.e., their coordinates are between 0 and 1. The function $h : [0,1] \times [0,1] \to \mathbf{R}$ is defined by $h(x) = h(x_1, x_2) = \min\{x_1, 1 - x_1\} \cdot \min\{x_2, 1 - x_2\}$.

The basic idea is to compute the convex hull of a small subset S of A such that $\text{conv}(S) = \text{conv}(A)$ with high probability. The value $h(x)$ will express whether x is likely to be an interior point of $\text{conv}(A)$. With increasing $h(x)$ the probability that x can be eliminated as a candidate for a vertex of the convex hull increases.

We will discard points based on this function in a first phase and compute the convex hull for the remaining points. It will turn out that we cannot guarantee that we have obtained $\text{conv}(A)$ this way. Therefore, in a second phase we have to check correctness and possibly correct the results.

The following is a sketch of our elimination type algorithm where CH is some algorithm to compute the convex hull of a set of points in the plane.

procedure elim_hull

Phase1

(1) Choose a suitable parameter α.

(2) Let $S_\alpha = \{a_i \in A \mid h(a_i) \leq \alpha\}$.

(3) Apply CH to compute the convex hull of S_α.

(4) Compute the minimum γ such that $\text{conv}(S_\alpha) \supseteq \{x \in [0,1] \times [0,1] \mid h(x) > \gamma\}$.

(5) If $\alpha \geq \gamma$ then STOP ($\text{conv}(S_\alpha) = \text{conv}(A)$), otherwise perform Phase 2.

Phase2

(6) Compute $\text{conv}(S_\gamma)$ where $S_\gamma = \{a_i \in A \mid h(a_i) \leq \gamma\}$. Now $\text{conv}(S_\gamma) = \text{conv}(A)$.

end of elim_hull

Figure 4.15 gives an illustration of the algorithm. The set S_α is given by the solid points ('•') and its convex hull by solid lines. The broken curve defines the set S_γ and the additional points to be considered in Phase 2 are shown as small circles ('∘'). The extreme points of the correct convex hull resulting from Phase 2 are obtained by adding one point in the north-east corner.
A detailed analysis shows that Step (4) can be performed in linear time. Therefore the worst case running time of this algorithm is given by the worst case running time of CH (independent of α).

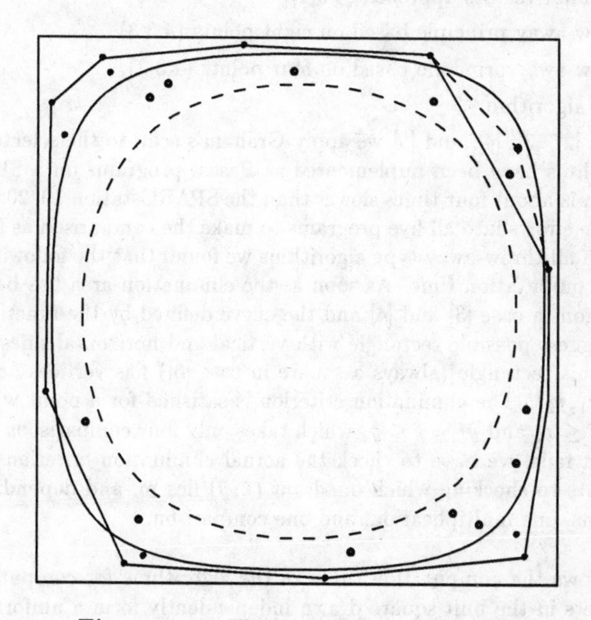

Figure 4.15 Illustration of the algorithm

The analysis of the average time complexity of this algorithm exhibits some interesting consequences. If α is chosen carefully, then in certain random models only very few points are contained in S_α and with very high probability Phase 2 is not needed. In particular, one can obtain a speed-up theorem for convex hull algorithms in the following sense.

Theorem 4.5 *Let A be a set of n random points generated independently from the uniform distribution on the unit square* $[0,1] \times [0,1]$. *For any algorithm CH with polynomial worst-case running time the two-phase method has linear expected running time.* □

For a detailed analysis of the algorithm and the proper choice of α as well as for a discussion of the computation of convex hulls for random points in the unit disk $D(0,1) = \{x = (x_1, x_2) \in \mathbf{R}^2 \mid x_1^2 + x_2^2 \leq 1\}$ we refer to BORGWARDT, GAFFKE, JÜNGER & REINELT (1991). Our approach can be generalized to higher dimensions. A complete coverage of the 3-dimensional case can be found in THIENEL (1991).

Assuming a uniform distribution of the n independent points over the unit square, linear expected time algorithms have been given by BENTLEY & SHAMOS (1978), AKL & TOUSSAINT (1978), DEVROYE (1980), DEVROYE & TOUSSAINT (1981), KIRKPATRICK & SEIDEL (1986), and GOLIN & SEDGEWICK (1988). For a survey on these and related subjects see LEE & PREPARATA (1984).

We have compared the practical behaviour of five linear expected time algorithms, namely

[1] the divide and conquer algorithm (4.3.2),

[2] the maximal vectors approach (4.3.4),

[3] the throw-away principle based on eight points (4.3.3),

[4] the throw-away principle based on four points (4.3.3),

[5] the new algorithm.

For algorithms [2], [3], [4], and [5] we apply Graham's scan to the selected points.

All five algorithms have been implemented as Pascal programs on a SUN SPARCstation SLC which is about four times slower than the SPARCstation 10/20. We have tried to put the same efforts into all five programs to make the comparison as fair as possible. For instance, in all throw-away type algorithms we found that the following trick helped to reduce the computation time. As soon as the elimination area has been determined (a closed polygon in case [3] and [4] and the curve defined by the function h in [5]) we inscribe the biggest possible rectangle with vertical and horizontal sides into this area. Assume that this rectangle (always a square in case [5]) has vertices (x_1, y_1), (x_2, y_1), (x_2, y_2) and (x_1, y_2). The elimination criterion is satisfied for a point with coordinates $(\overline{x}, \overline{y})$ if $x_1 \leq \overline{x} \leq x_2$ and $y_1 \leq \overline{y} \leq y_2$ which takes only four comparisons to check. Only if this criterion fails, we have to check the actual elimination criterion which, e.g., in case [5] amounts to checking which quadrant $(\overline{x}, \overline{y})$ lies in, and depending on this, up to two additions, one multiplication, and one comparison.

Figure 4.16 shows the computation times of the algorithms for computing the convex hull of point sets in the unit square drawn independently form a uniform distribution. The curves are based on 10 sample problems for each problem size $n = 1\,000, 2\,000, \ldots,$ $10\,000, 20\,000, \ldots, 100\,000, 200\,000, \ldots, 1\,000\,000$. In our opinion, these curves should be interpreted as follows. When doing practical computations, the throw-away principle is superior compared to the divide and conquer algorithms. The four point method is slightly better than the eight point method.

For our experiments with practical traveling salesman problems in the plane we have coded the algorithm as follows. The points are mapped to the unit square by horizontal

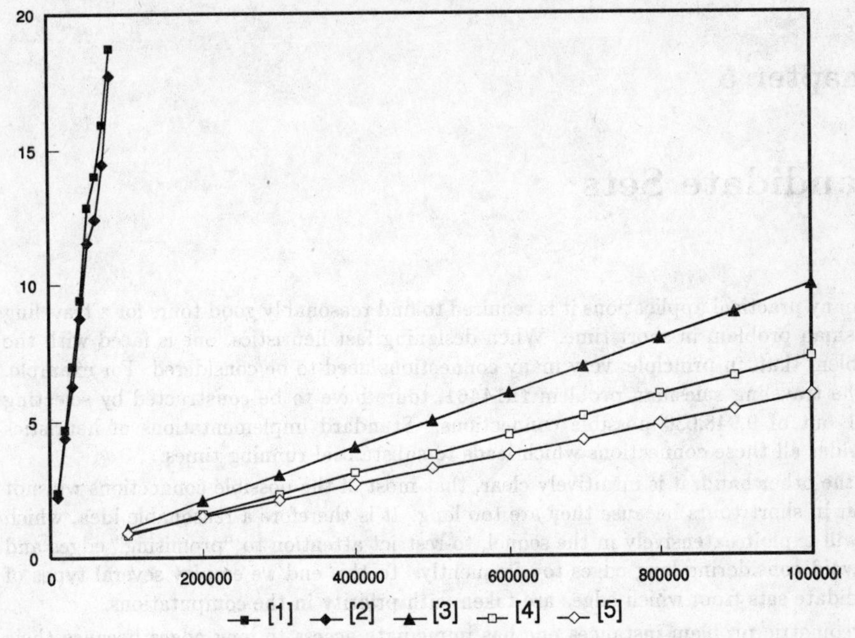

Figure 4.16 Comparison of convex hull algorithms

and vertical scaling and appropriate offsets. This is done in such a way that each of the four sides of the square contains at least one problem point. Then the elimination algorithm is performed using $\alpha = 4\log n/n$. However, in the range of problem sizes we had available, the new algorithm does not pay off since the convex hull of some ten thousands of points can be computed in so little time that it is negligibly small compared to the other algorithms employed for problem solving. Therefore we do not report computing times for TSPLIB problems here.

Chapter 5

Candidate Sets

In many practical applications it is required to find reasonably good tours for a traveling salesman problem in short time. When designing fast heuristics, one is faced with the problem that, in principle, very many connections need to be considered. For example, in the traveling salesman problem fnl4461, tours have to be constructed by selecting 4461 out of 9,948,030 possible connections. Standard implementations of heuristics consider all these connections which leads to substantial running times.

On the other hand, it is intuitively clear, that most of the possible connections will not occur in short tours because they are too long. It is therefore a reasonable idea, which we will exploit extensively in the sequel, to restrict attention to "promising" edges and to avoid considering long edges too frequently. To this end we employ several types of candidate sets from which edges are taken with priority in the computations.

In geometric problem instances one has immediate access to long edges because their length is related to the location of the points. In general, for problems given by a distance matrix, already time $\Omega(n^2)$ has to be spent to scan all edge lengths. We will discuss three types of candidate sets in this chapter. The first one is applicable in general, but can be computed very fast for geometric instances. The other two sets can only be computed for geometric instances.

5.1 Nearest Neighbors

It can be observed that most of the edges in good or optimal tours connect nodes to near neighbors. For a TSP on n nodes and $k \geq 1$, we define the corresponding k **nearest neighbor subgraph** $G_k = (V, E)$ by setting

$$\begin{aligned} V &= \{1, 2 \ldots, n\}, \\ E &= \{uv \mid v \text{ is among the } k \text{ nearest neighbors of } u\}. \end{aligned}$$

For example, an optimal solution for the problem pr2392 can be found within the 8 nearest neighbor subgraph and for pcb442 even within the subgraph of the 6 nearest neighbors.

Figure 5.1 shows the 10 nearest neighbor subgraph for the problem u159. This subgraph contains an optimal tour.

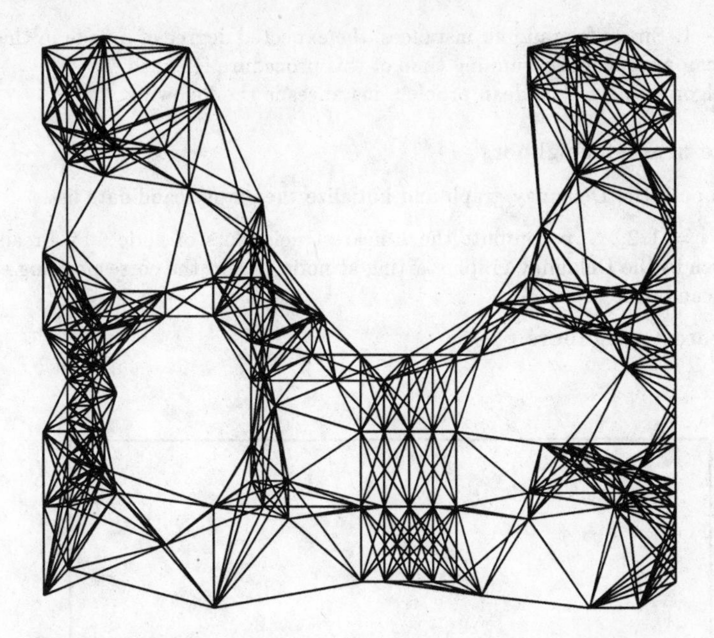

Figure 5.1 The 10 nearest neighbor subgraph for **u159**

A straightforward computation of the k nearest neighbors by enumeration takes time $\Omega(n^2)$ for fixed k. The following proposition shows that for Euclidean problem instances we can exploit the Delaunay graph for nearest neighbor computations. The discussion applies to other metrics as well.

Proposition 5.1 *Let P_i and P_j be two generators.*

(i) If the straight line connecting P_i and P_j intersects the interior of the Voronoi region of a generator P_l different from P_i and P_j then $d(P_i, P_l) < d(P_i, P_j)$.

(ii) If the smallest number of edges on a path from P_i to P_j in the Delaunay graph is k then there exist at least $k-1$ generators P_l different from P_i and P_j with $d(P_i, P_l) < d(P_i, P_j)$.

Proof. For part (i) suppose that the line intersects the boundary of $\mathrm{VR}(P_l)$ in the points T_1 and T_2 where w. l. o. g. $d(P_i, T_1) < d(P_i, T_2)$. By definition we have $d(P_l, T_1) \leq d(P_i, T_1)$ and $d(P_l, T_2) \leq d(P_j, T_2)$. Since P_i, P_l, and T_1 are distinct we obtain

$$\begin{aligned} d(P_i, P_l) &\leq d(P_i, T_1) + d(P_l, T_1) \\ &< d(P_i, T_1) + d(T_1, P_j) \\ &= d(P_i, P_j). \end{aligned}$$

Part (ii) is an immediate corollary. □

Therefore, to compute the k nearest neighbors of some generator P_i we only have to examine generators which are connected to P_i in the Delaunay graph by a path of length

at most $k-1$. Since, for random instances, the expected degree of a node in this graph is six, we can expect linear running time of this procedure for fixed k.
The fast algorithm for Euclidean problem instances is the following.

procedure nearest_neighbors

(1) Compute the Delaunay graph and initialize the empty candidate list.

(2) For $i = 1, 2, \ldots, n$ compute the k nearest neighbors of node i by breadth-first search in the Delaunay graph starting at node i. Add the corresponding edges to the candidate set.

end of nearest_neighbors

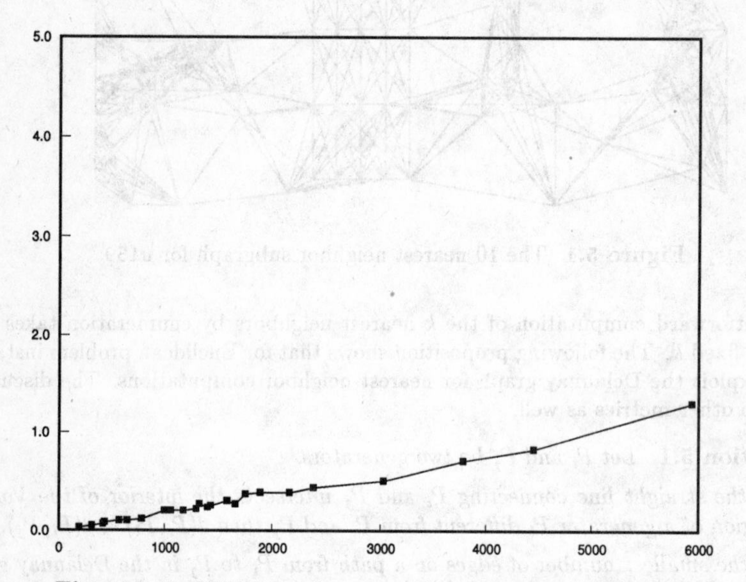

Figure 5.2 CPU times for computing 10 nearest neighbor graph

Figures 5.2 and 5.3 show the running times for the 10 nearest neighbor computations for our set of sample problems as well as the number of edges in the resulting candidate sets. A good approximation for the cardinality of the 10 nearest neighbor candidate set is $6n$.
It is interesting to note that computation of the 10 nearest neighbor set for problem pr2392 takes 22.3 seconds if the trivial algorithm is used. If the complete distance matrix is stored this time reduces to 18.8 seconds which is still substantially more than the 0.5 seconds needed with the Delaunay graph. It should be kept in mind that we do not use complete distance matrices in this tract, but that much CPU time can be gained if they are available.

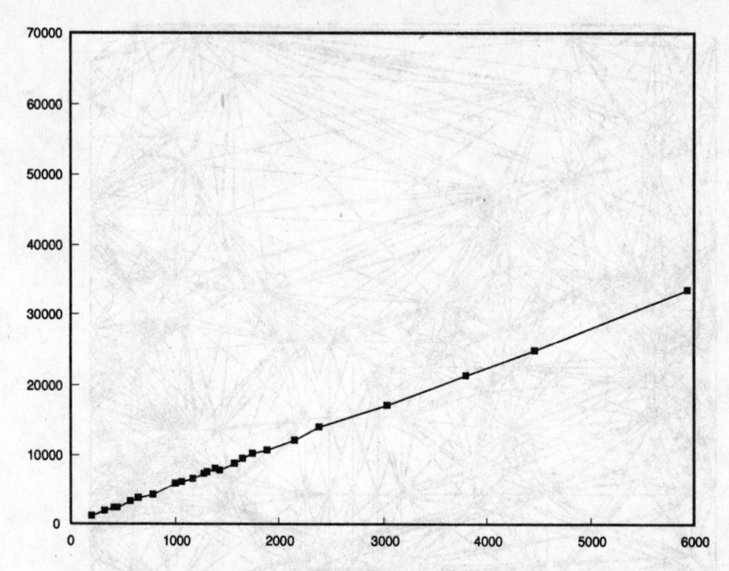

Figure 5.3 Number of edges of the 10 nearest neighbor graphs

There are further approaches for an efficient computation of nearest neighbors for geometric problems instances, e.g., probabilistic algorithms and algorithms based on k-d-trees (BENTLEY (1990)).

5.2 Candidates Based on the Delaunay Graph

In particular if point sets exhibit several clusters, the k nearest neighbor subgraph is not connected and many edges to form good tours are missing. Here the Delaunay graph should help since it contains important connections between the clusters.
Though it may seem to be true at a first glance, the Delaunay graph itself does not necessarily contain a Hamiltonian tour. For example, in the case where all points are on a line the Delaunay graph is just a path. There are also examples where the Delaunay graph is a triangulation, but does not contain a tour or even a Hamiltonian path (DILLENCOURT (1987a,1987b)).
First experiments have indicated that the Delaunay graph provides a candidate set too small. We therefore decided to augment it using transitive relations in the following way. If the edges $\{i, j\}$ and $\{j, k\}$ are contained in the Delaunay graph then we also add edge $\{i, k\}$ to the candidate set. We call this set **Delaunay candidate set**. The cardinality of this set can be quite large. For example, if $n - 1$ generators are located on a circle and one generator is at the center of this circle then the Delaunay candidate set is the complete graph on n nodes. Figure 5.4 shows the Delaunay candidate set for problem u159.

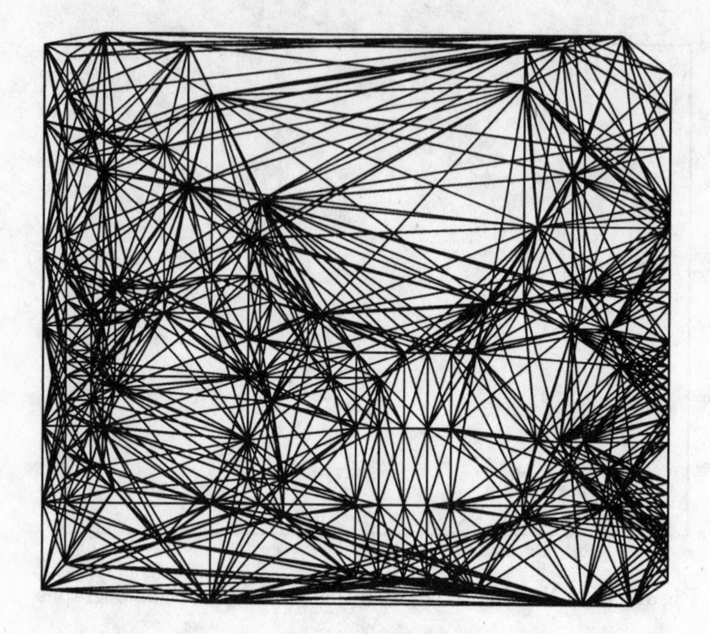

Figure 5.4 The Delaunay candidate set for **u159**

The following procedure computes the Delaunay candidate set.

procedure Delaunay_candidates

(1) Compute the Delaunay graph and initialize the candidate set with the edges of the Delaunay graph.

(2) For every node $i = 1, 2, \ldots, n$ do

(2.1) For every two nodes j and k adjacent to i in the Delaunay graph add edge $\{j, k\}$ to the candidate set if it was not a candidate edge before.

end of Delaunay_candidates

For random points in the plane, empirical observations show that we can expect about $9n$ to $10n$ edges in this candidate set.
Figure 5.4 illustrates that the candidate set is rather dense. Therefore we have to expect more than $9n$ edges in this candidate set for practical problem instances. Furthermore, due to long edges in the Delaunay graph, the candidate set may contain many long edges. To avoid long edges, we usually first run a fast heuristic to compute an initial tour (as the space filling curves heuristic described in Chapter 8) and then eliminate all edges from the candidate set that are longer than the longest edge in this tour. For dense point sets most long edges will be eliminated, for clustered point sets the edges connecting the clusters will be kept. In general, however, elimination of long edges is not critical for the performance of our heuristics.

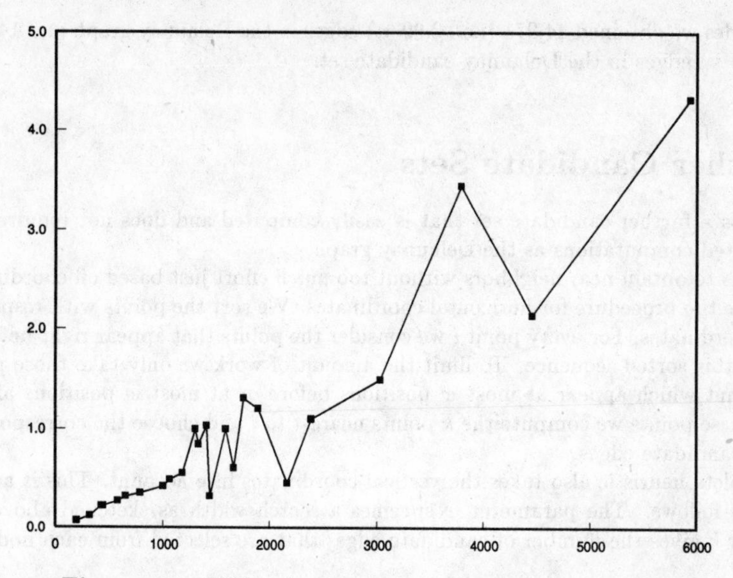

Figure 5.5 CPU times for computing Delaunay candidate sets

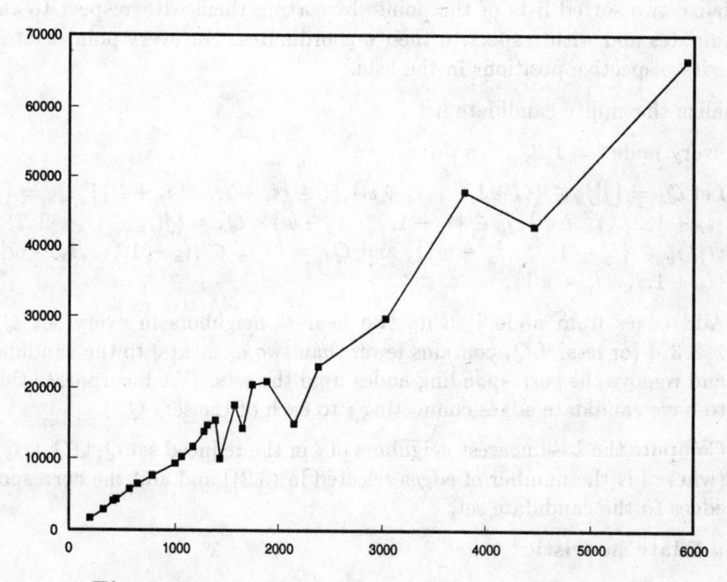

Figure 5.6 Number of edges of Delaunay candidate sets

Figures 5.5 and 5.6 give the CPU times necessary for computing the Delaunay candidate set and the cardinality of this set. CPU time and size of the candidate set depend highly on the point configuration and not only on the problem size. For a random problem on

15,000 nodes we obtained 44,971 (i.e., $2.99 \cdot n$) edges in the Delaunay graph and 147,845 (i.e., $9.86 \cdot n$) edges in the Delaunay candidate set.

5.3 Other Candidate Sets

We discuss a further candidate set that is easily computed and does not require such sophisticated computations as the Delaunay graph.
The idea is to obtain near neighbors without too much effort just based on coordinates. We outline the procedure for horizontal coordinates. We sort the points with respect to their x-coordinates. For every point i we consider the points that appear right before or after i in this sorted sequence. To limit the amount of work we only take those points into account which appear at most w positions before or at most w positions after i. Among these points we compute the k points nearest to i and choose the corresponding edges as candidate edges.
The complete heuristic also takes the vertical coordinates into account. This is accomplished as follows. The parameter w specifies a search width as sketched above, the parameter k gives the number of candidate edges that are selected from each node.

procedure candidate_heuristic

(1) Initialize two sorted lists of the points by sorting them with respect to their x-coordinates and with respect to their y-coordinates. For every point i let i_x and i_y be its respective positions in the lists.

(2) Initialize the empty candidate list.

(3) For every node $i = 1, 2, \ldots, n$ do

(3.1) Let $Q_1 = \{j | j_x \in \{i_x+1, \ldots, i_x+w\}, j_y \in \{i_y+1, \ldots, i_y+w\}\}$, $Q_2 = \{j | j_x \in \{i_x+1, \ldots, i_x+w\}, j_y \in \{i_y-1, \ldots, i_y-w\}\}$, $Q_3 = \{j | j_x \in \{i_x-1, \ldots, i_x-w\}, j_y \in \{i_y-1, \ldots, i_y-w\}\}$, and $Q_4 = \{j | j_x \in \{i_x-1, \ldots, i_x-w\}, j_y \in \{i_y+1, \ldots, i_y+w\}\}$.

(3.2) Add edges from node i to its two nearest neighbors in every set $Q_j, j = 1, 2, 3, 4$ (or less, if Q_j contains fewer than two elements) to the candidate set and remove the corresponding nodes from the sets. (We incorporate this step to have candidate edges connecting i to each of the sets Q_j.)

(3.3) Compute the $k - l$ nearest neighbors of i in the reduced set $Q_1 \cup Q_2 \cup Q_3 \cup Q_4$ (where l is the number of edges selected in (3.2)) and add the corresponding edges to the candidate set.

end of candidate_heuristic

Figure 5.7 shows the candidate set obtained with this heuristic for problem instance **u159** (parameters were $k = 10$ and $w = 20$). It contains 94% of the edges of the 10 nearest neighbor graph. Because of pathological conditions at the border of point sets we may also incur a number of long edges in this heuristic. These could be eliminated as above, but keeping them has no significant effects as our experiments showed.

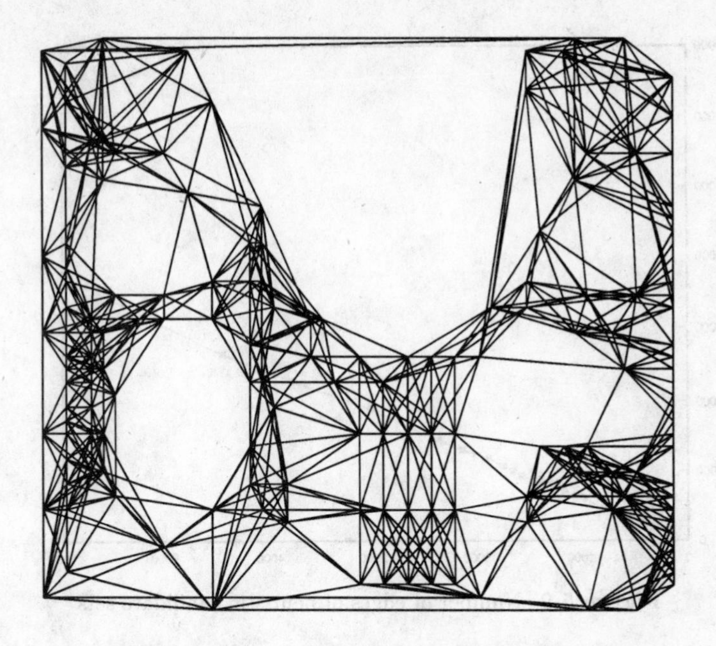

Figure 5.7 Result of candidate heuristic for u159

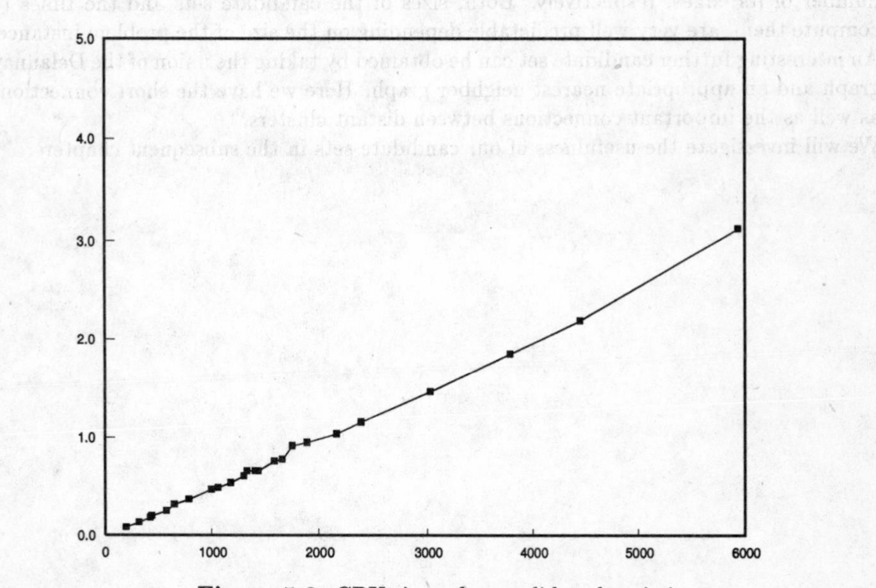

Figure 5.8 CPU times for candidate heuristic

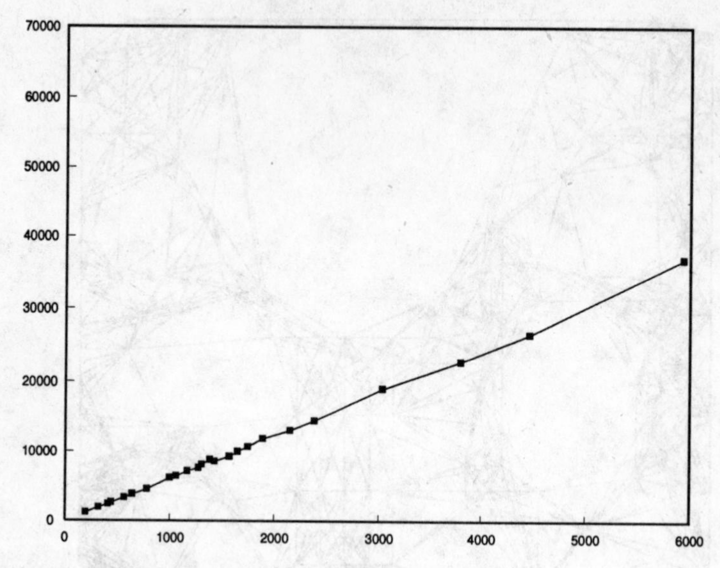

Figure 5.9 Number of edges of heuristic candidate sets

Figures 5.8 and 5.9 display the CPU times for computing this candidate set and the number of its edges, respectively. Both, sizes of the candidate sets and the times to compute them, are very well predictable depending on the size of the problem instance. An interesting further candidate set can be obtained by taking the union of the Delaunay graph and an appropriate nearest neighbor graph. Here we have the short connections as well as the important connections between distant clusters.

We will investigate the usefulness of our candidate sets in the subsequent chapters.

Chapter 6

Construction Heuristics

Starting with this chapter we will now consider computational aspects of the traveling salesman problem. For the beginning we shall consider pure **construction procedures**, i.e., heuristics that determine a tour according to some construction rule, but do not try to improve upon this tour. In other words, a tour is successively built and parts already built remain in a certain sense unchanged throughout the algorithm.
Many of the construction heuristics presented here are known and computational results are available (GOLDEN & STEWART (1985), ARTHUR & FRENDEWAY (1985), JOHNSON (1990), BENTLEY (1992)) We include them for the sake of completeness of this tract and for having a reference to be compared with other algorithms on our sample problem instances. Moreover, most evaluations of heuristics performed in the literature lack from the fact that either fairly small problem instances or only random instances were examined.
The following types of algorithms will be discussed:

- nearest neighbor heuristics,
- insertion heuristics,
- heuristics based on spanning trees, and
- savings heuristics.

This does not cover by far all the approaches that have been proposed. But we think that the ideas presented here provide the reader with the basic principles that can also be adapted to other combinatorial optimization problems.
For the following, we will always assume that we are given the complete undirected graph K_n with edge weights c_{uv} for every pair u and v of nodes. For ease of notation we will denote the node set by V and assume that $V = \{1, 2, \ldots, n\}$. The question to be addressed is to find good Hamiltonian tours in this graph.

6.1 Nearest Neighbor Heuristics

This heuristic for constructing a traveling salesman tour is near at hand. The salesman starts at some city and then visits the city nearest to the starting city. From there he visits the nearest city that was not visited so far, etc., until all cities are visited, and the salesman returns to the start,

6.1.1 The Standard Version

Formulated as an algorithm we obtain the following procedure.

procedure nearest_neighbor

(1) Select an arbitrary node j, set $l = j$ and $T = \{1, 2, \ldots, n\} \setminus \{j\}$.

(2) As long as $T \neq \emptyset$ do the following.

(2.1) Let $j \in T$ such that $c_{lj} = \min\{c_{li} \mid i \in T\}$.

(2.2) Connect l to j and set $T = T \setminus \{j\}$ and $l = j$.

(3) Connect l to the first node (selected in Step (1)) to form a tour.

end of nearest_neighbor

This procedure runs in time $\Omega(n^2)$. A possible variation of the standard nearest neighbor heuristic is the **double-sided nearest neighbor heuristic** where the current path can be extended from both of its endnodes.
No constant worst case performance guarantee can be given, since the following theorem due to ROSENKRANTZ, STEARNS & LEWIS (1977) holds.

Theorem 6.1 *For every $r > 1$ and arbitrarily large n there exists a TSP instance on n cities such that the nearest neighbor tour is at least r times as long as an optimal tour.*

□

In addition, ROSENKRANTZ, STEARNS AND LEWIS (1977) show that for arbitrarily large n there exist TSP instances on n nodes such that the nearest neighbor solution is $\Theta(\log n)$ times as long as an optimal Hamiltonian cycle. This results still holds if the triangle inequality is satisfied. Therefore it also applies to metric problem instances.

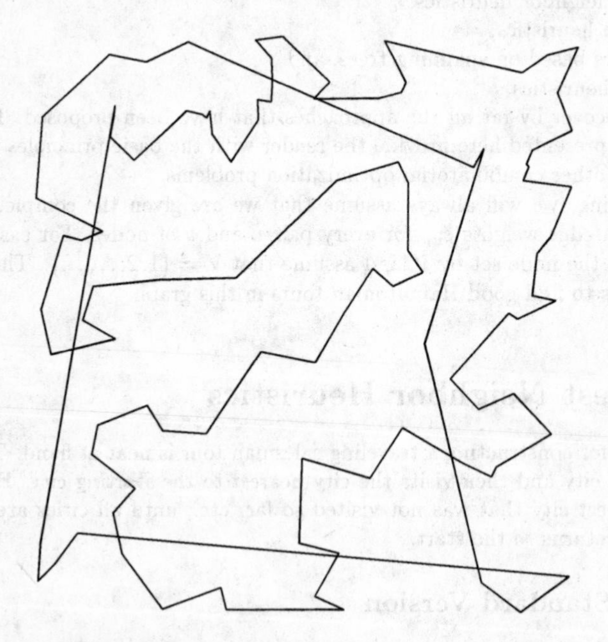

Figure 6.1 A nearest neighbor tour for rd100

If one displays nearest neighbor tours one realizes the reason for their poor performance. The procedure proceeds very well and produces connections with short edges in the beginning. But, as can be seen from a graphics display, several cities are "forgotten" during the course of the algorithm. They have to be inserted at high cost in the end. Figure 6.1 shows a typical nearest neighbor tour.

Though usually rather bad, nearest neighbor tours have the advantage that they only contain a few severe mistakes, but there are long segments connecting nodes with short edges. Therefore, such tours can serve as good starting tours for subsequently performed improvement methods and it is reasonable to put some effort in designing heuristics that are based on the nearest neighbor principle. We will comment on improvement methods in the next chapter. The standard procedure itself is easily implemented with a few lines of code. But, since running time is quadratic, we describe some variants to speed up and/or improve the standard nearest neighbor search.

6.1.2 Exploiting the Delaunay Graph

We have seen in Chapter 5 that the Delaunay graph can be used to speed up nearest neighbor computations. We can apply these results here, too. Namely, when searching the nearest neighbor of node l in step (2.1) among the nodes which are not yet contained in the partial tour, we can use the principle of section 5.1 to generate the k-th nearest neighbor of l for $k = 1, 2, \ldots, n$ until a node is found that is not yet connected. Due to the properties of the Delaunay graph we should find this neighbor examining only a few edges of the graph in the neighborhood of l. Since in the last steps of the algorithm we have to collect the forgotten nodes (which are far away from the current node) it makes no sense to use the Delaunay graph any further. So, for connecting the final nodes we just use a simple enumeration procedure.

We have conducted several experiments to see how many neighbors of the current node are examined and to what depth the breadth-first search to find the nearest neighbor is performed.

Figure 6.2 shows the average depth in the breadth-first search tree needed to find the next neighbor. The average depth varies between 3 and 5 for real-world problems and is about 3.5 for random problems.

Figure 6.3 displays the average number of nodes examined to find the next neighbor, which is then also the number of necessary distance evaluations per node. Here real-world problems behave better than random problems.

Furthermore, we examined how search depth and number of neighbors to be examined develop during the heuristic. Figures 6.4 and 6.5 depict the search depth and the number of examined nodes, respectively, obtained during the execution of the nearest neighbor heuristic on the problem pr2392.

We see that at the majority of nodes next neighbors can indeed be found in the local neighborhood with a search depth below five in most cases. Only sometimes a large part of the Delaunay graph has to be explored to find the next node. Note that we do not use the Delaunay strategy for inserting the final 100 nodes, since the overhead for identifying the nearest neighbor increases significantly at the end of the procedure.

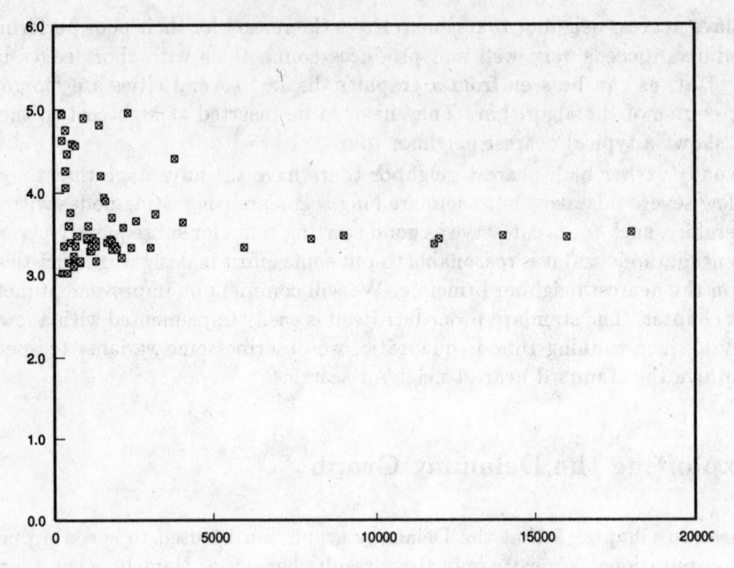

Figure 6.2 Average search depth

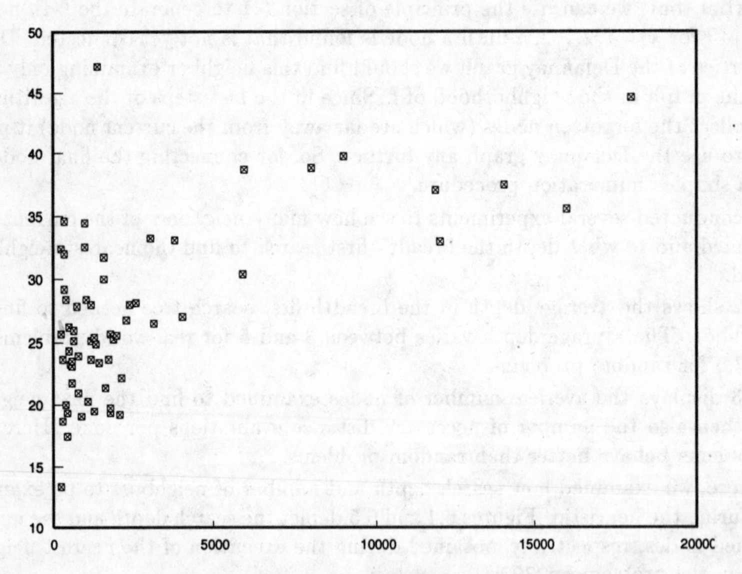

Figure 6.3 Average number of examined nodes

The worst case running time of this implementation is still quadratic. The running times for the sample problems will be given below together with the running times for all other variants.

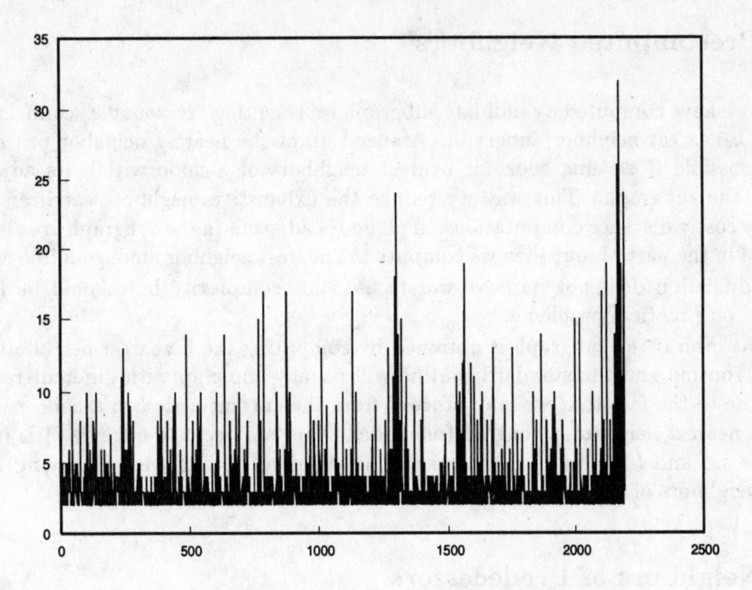

Figure 6.4 Search depth for pr2392

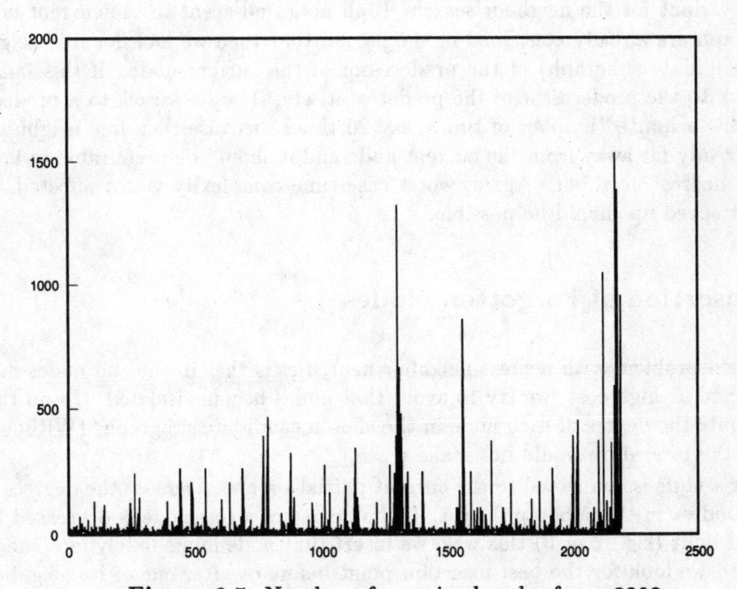

Figure 6.5 Number of examined nodes for pr2392

6.1.3 Precomputed Neighbors

Suppose we have computed a candidate subgraph representing "reasonable" connections, e.g., the k nearest neighbor subgraph. A speed up of the nearest neighbor procedure is then possible if we first look for nearest neighbors of a node within its adjacent nodes in the subgraph. This way we reduce the exhaustive neighbor search and the necessary costly distance computations. If all nodes adjacent in the subgraph are already contained in the partial tour then we compute the nearest neighbor among all free nodes. This modification does not improve worst case time complexity but should be faster when run on practical problems.
Note, that even if the subgraph is obtained by computing the k nearest neighbors this modified routine and the standard routine will usually come up with different results. This is due to the fact that we may proceed from the current node l to a node j which is not its nearest neighbor among all free nodes. This can occur, if edge $\{l, j\}$ is in the candidate set and l is among the k nearest neighbors of j, but j is not among the k nearest neighbors of l.

6.1.4 Neighbors of Predecessors

In this modification we also use a precomputed set of candidate edges but apply the following variant for the neighbor search. If all nodes adjacent to the current node in the subgraph are already contained in the partial tour then we look for free neighbors (in the candidate subgraph) of the predecessor of the current node. If this fails too, we go back to the predecessor of the predecessor, etc. The backtrack to a predecessor is only done a limited number of times, say 20 times, because then free neighbors are usually already far away from the current node and it should be preferrable to look for the exact nearest neighbor. Again, worst case time complexity is not affected, but a significant speed up should be possible.

6.1.5 Insertion of Forgotten Nodes

As the main problem with nearest neighbor heuristics is that in the end nodes have to be connected at high cost, we try to avoid that nodes become isolated. To do this we first compute the degree of each node in the chosen candidate subgraph. (Without this subgraph the procedure would not make sense.)
Whenever a node is connected to the current partial tour we decrease the degrees of its adjacent nodes (in the subgraph) by 1. If the degree of a free node is decreased below a specified limit (e.g., 2 or 3) this way, we insert that node immediately into the path. To this end we look for the best insertion point before or after one of its neighbors in the candidate subgraph. This way more nodes of degree less than or equal to the given limit may occur which are also inserted rightaway. The selection of the next nodes in the algorithm is accomplished as in variant 6.1.3. The worst case time complexity is still quadratic.

6.1.6 Using Rotation Operations

The idea of this heuristic is to try to grow the tour within the candidate subgraph. If at the current node the tour cannot be extended using candidate edges it is tried to perform a sequence of rotation operations. Such an operation introduces an unused subgraph edge from the current node. Since this results in a cycle with the partial tour, another edge has to be eliminated and we obtain a new last node from which we can try to extend the path.

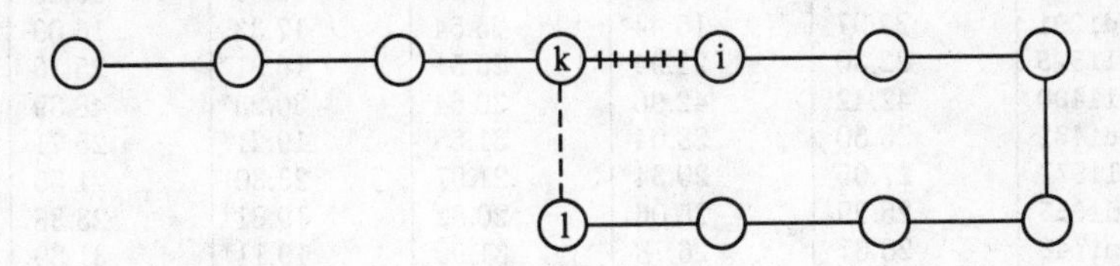

Figure 6.6 A rotation operation

Figure 6.6 depicts a rotation operation. If at the current last node l the path cannot be extended within the subgraph we try to use the subgraph edge $\{l, k\}$. To break the resulting cycle we delete the edge $\{k, i\}$ and try to extend the path now starting at i. A sequence of rotations can be performed if also the extension from i fails. If the tour cannot be extended this way using only subgraph edges a neighbor search like in one of the previous variants can be performed.

6.1.7 Comparison of Variants

We compare the variants with the standard heuristic implemented according to 6.1.2. It is clear that the tours produced by the algorithms heavily depend on the choice of the starting point. Since we cannot look for the best starting point we always chose to start with node $\lfloor \frac{n}{2} \rfloor$, thus giving an unbiased starting node to the heuristics. The chosen candidate subgraph was the 10 nearest neighbor subgraph for variants 6.1.3 through 6.1.6.

In variant 6.1.4 we examined at most 20 predecessors to extend the path. In variant 6.1.5 forgotten nodes were inserted as soon as they were merely connected to at most three free nodes. In variant 6.1.6 a sequence of rotations was limited to be composed of at most five single rotations.

Table 6.7 shows the tour lengths (given as deviation in percent from the best known lower bounds) obtained by applying the different procedures to our set of test problems. Variants 6.1.2 through 6.1.6 are denoted by Variant 1 through Variant 5 in this table. The best solution found for every problem instance is marked with a '*'.

The results strongly support variant 6.1.5 which avoids adding too many isolated nodes in the end. Usually, this decreases the tour length considerably. The quality of the solutions can be expected to be in the range of 15% to 25% above optimality. In JOHNSON (1990) an average excess of 24% over an approximation of the Held-Karp lower bound (see Chapter 10) is reported for randomly generated problems.

Problem	Variant 1	Variant 2	Variant 3	Variant 4	Variant 5
d198	25.79	16.51*	25.86	20.88	28.09
lin318	26.85	22.52	33.06	15.90*	27.84
fl417	21.28	17.84*	36.42	25.36	31.80
pcb442	21.36	22.91	18.63	13.51*	27.15
u574	29.60	21.11	25.90	18.67*	29.16
p654	31.02	18.75*	25.80	25.98	24.43
rat783	27.13	24.76	27.98	18.86*	27.02
pr1002	24.35	25.18	28.96	18.16*	27.97
u1060	30.43	27.14	27.33	24.14*	29.42
pcb1173	28.18	27.69	27.30	18.09*	29.85
d1291	22.97	15.44*	25.54	17.33	16.09
rl1323	22.30	21.94	25.51	16.81*	25.10
fl1400	42.42	42.96	30.64	30.23*	48.39
u1432	25.50	28.04	31.55	19.21*	25.71
fl1577	27.65	20.34*	21.97	23.30	21.55
d1655	25.99	25.06	20.82	19.81*	23.38
vm1748	25.67	26.78	31.90	19.11*	31.89
rl1889	28.37	25.54	23.85*	25.62	25.41
u2152	25.80	25.85	23.22	18.97*	23.14
pr2392	24.96	27.23	26.09	22.68*	26.20
pcb3038	23.63	23.53	28.39	19.16*	26.92
fl3795	24.44	25.92	32.85	20.09*	35.24
fnl4461	25.31	23.24	26.98	19.53*	24.99
rl5934	22.93	23.41	24.77	18.75*	24.38
Average	26.27	26.58	27.04	21.45	28.28

Table 6.7 Results of nearest neighbor variants

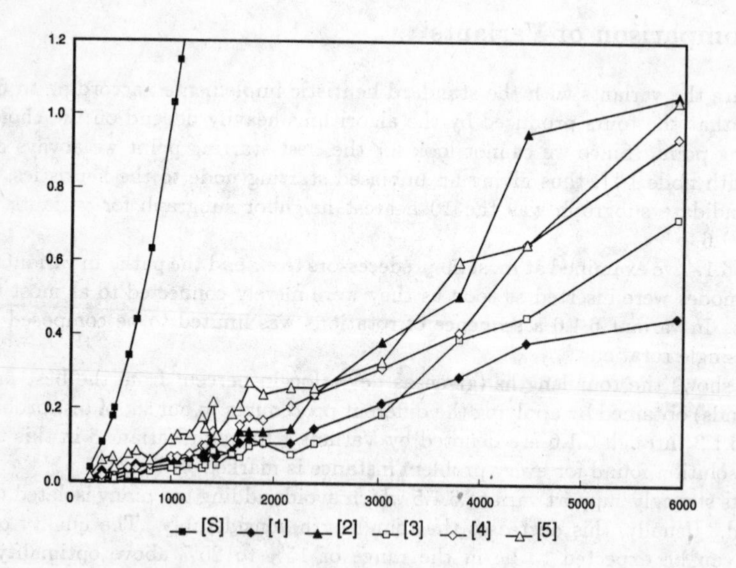

Figure 6.8 CPU times for nearest neighbor variants

CPU times for the complete set of instances are shown in Figure 6.8. The running times for the variants do not include the time to set up the Delaunay graph or the 10 nearest neighbor subgraph. These times were given in Chapters 4 and 5. To document the speed up obtained by using the different variants we have also included the running time for the standard implementation. Variant i is indicated by the number [i], the standard implementation is indicated by [S].

Figure 6.8 clearly visualizes that already simple algorithms of quadratic time complexity take quite some time when applied to large problems. The comparison of the variants gives a clear picture. All the variants are much faster than the standard nearest neighbor algorithm (even if the preprocessing times would be included). When considering larger random problems (results are not displayed here), variants 2, 3, and 4 seem to exhibit a quadratic component in their running time while variants 1 and 5 seem to have subquadratic running times.

6.1.8 Stability of Nearest Neighbor Heuristics

Since we have performed only one run of each heuristic for every sample problem (starting with node $\lfloor \frac{n}{2} \rfloor$) we cannot be absolutely sure that Table 6.7 gives a correct assessment of the five heuristics. We have therefore examined the average quality of each variant for three sample problems. To this end we have performed each heuristic for every starting node $l = 1, 2, \ldots, n$.

Table 6.9 shows the results. Each line corresponds to one variant and gives (in that sequence) the length of the best, resp. worst tour, the average tour length obtained, the span between best and worst tour (i.e., worst quality − best quality), and the standard deviation.

Variant	Minimum	Maximum	Average	Span	Deviation
lin318					
1	16.89	39.82	24.89	22.92	2.99
2	18.58	35.96	25.88	17.38	2.96
3	20.86	37.66	27.12	16.80	3.65
4	12.96	29.24	20.25	16.28	3.00
5	21.54	36.94	29.37	15.40	3.07
pcb442					
1	18.01	38.15	29.66	20.14	3.63
2	16.60	32.23	22.11	15.63	2.84
3	17.19	34.31	26.06	17.12	3.44
4	11.77	30.03	16.63	18.27	2.74
5	15.67	36.30	25.81	20.63	3.26
u1060					
1	22.33	36.51	26.11	14.17	1.93
2	21.25	39.19	30.05	17.94	3.45
3	24.18	37.99	28.89	13.81	2.21
4	19.04	28.46	22.95	9.42	1.70
5	24.98	37.44	30.03	12.46	2.29

Table 6.9 Sensitivity analysis for nearest neighbor variants

The results verify that insertion of forgotten neighbors leads to the best results. The average quality of the tours obtained this way is substantially better than for the other four variants. The other variants perform more or less the same. The span is considerable, the quality of the tours strongly depends on the choice of the starting node.

6.2 Insertion Heuristics

A further intuitive approach is to start with tours on small subsets (including trivial "tours" on one or two nodes) and then extend these tours by inserting the remaining nodes. This principle is realized by the following procedure.

procedure insertion

(1) Select a starting tour through k nodes $v_1, v_2, \ldots, v_k$ ($k \geq 1$) and set $W = V \setminus \{v_1, v_2, \ldots, v_k\}$.

(2) As long as $W \neq \emptyset$ do the following.

(2.1) Select a node $j \in W$ according to some criterion.

(2.2) Insert j at some position in the tour and set $W = W \setminus \{j\}$.

end of insertion

Using this principle a tour is built containing more and more nodes of the problem until all nodes are inserted and the final Hamiltonian tour is found.

6.2.1 Standard Versions

Of course, there are several possibilities for implementing such an insertion scheme. The main difference is the determination of the order in which the nodes are inserted. The starting tour is usually just some tour on three nodes or, an edge ($k = 2$), or even a loop containing only one node ($k = 1$). We will consider also another type of starting tour below. The selected node to be inserted is usually inserted into the tour at the point causing shortest increase in the length of the tour.

We say that a node is a **tour node** if it is already contained in the partial tour. For $j \in W$ we define $d_{\min}(j) = \min\{c_{ij} \mid i \in V \setminus W\}$, $d_{\max}(j) = \max\{c_{ij} \mid i \in V \setminus W\}$, and $s(j) = \sum_{i \in V \setminus W} c_{ij}$.

The following possibilities for extending the current tour are considered.

6.2.1.1 Nearest Insertion

Insert the node that has the shortest distance to a tour node, i.e., select j with $d_{\min}(j) = \min\{d_{\min}(l) \mid l \in W\}$. □

6.2.1.2 Farthest Insertion 1

Insert the node whose minimal distance to a tour node is maximal, i.e., select j with $d_{\min}(j) = \max\{d_{\min}(l) \mid l \in W\}$. □

6.2.1.3 Farthest Insertion 2

Insert the node that has the farthest distance to a tour node, i.e., select j with $d_{\max}(j) = \max\{d_{\max}(l) \mid l \in W\}$. □

6.2.1.4 Farthest Insertion 3

Insert the node whose maximal distance to a tour node is minimal, i.e., select j with $d_{\max}(j) = \min\{d_{\max}(l) \mid l \in W\}$. □

6.2.1.5 Cheapest Insertion 1

Among all nodes not inserted so far, choose a node whose insertion causes the lowest increase in the length of the tour. I.e., among all nodes not inserted so far, choose a node which can be inserted causing the lowest increase in the length of the tour. □

6.2.1.6 Cheapest Insertion 2

In the cheapest insertion heuristic, we have to know for every node not in the tour its cheapest insertion point. Update of this information is expensive (see below). In this variant, we only perform a partial update of the best insertion points in the following sense. Suppose node j has just been inserted into the partial tour. This may have the effect that the best insertion point changes for a non-tour node, say l. Now, we do not consider all possibilities to insert l, but only insertions before or after j and a limited number k of j's successors and predecessors. This has the consequence, that, for some nodes, not necessarily the best insertion point is determined. □

6.2.1.7 Random Insertion

Select the node to be inserted at random. □

6.2.1.8 Largest Sum Insertion

Insert the node whose sum of distances to tour nodes is maximal, i.e., select j with $s(j) = \max\{s(l) \mid l \in W\}$. This is equivalent to choosing the node with maximal average distance to tour nodes. □

6.2.1.9 Smallest Sum Insertion

Insert the node whose sum of distances to tour nodes is minimal, i.e., select j with $s(j) = \min\{s(l) \mid l \in W\}$. This is equivalent to choosing the node with minimal average distance to tour nodes. □

There are also variants of these ideas where the node selected is not inserted at cheapest insertion cost but as a neighbor of that tour node that is nearest to it. These variants are usually named "addition" instead of insertion. BENTLEY (1992) reports that the results are slightly inferior.

All heuristics except for cheapest Insertion have running time $O(n^2)$. Cheapest Insertion can be implemented to be executed in time $O(n^2 \log n)$ by storing for each external node a heap based on the insertion cost at the possible insertion points. Because this procedure requires $O(n^2)$ space it cannot be used for large problem instances. The fast version of cheapest insertion runs in time $O(n^2)$ because of the limited update.

We give an illustration of insertion principles in Figure 6.10 for a Euclidean problem instance. In the next step nearest insertion adds node i, farthest insertion adds node j, and cheapest insertion adds node k to the tour.

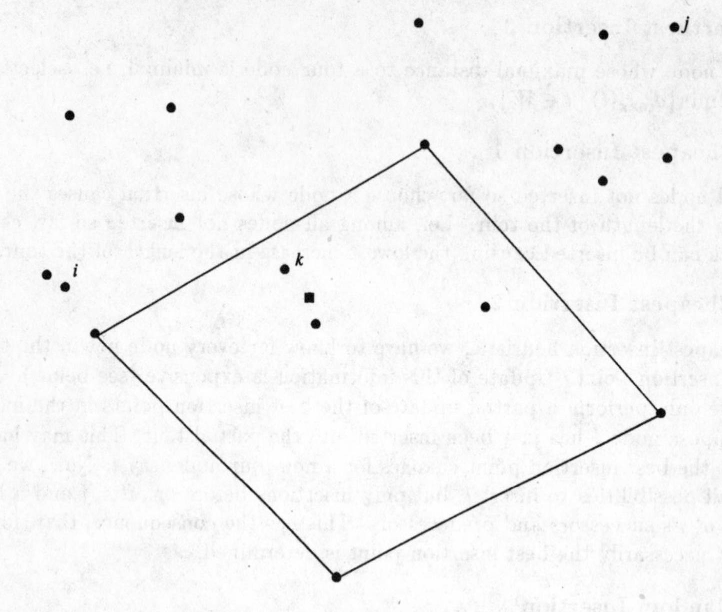

Figure 6.10 Illustration of insertion heuristics

Nearest insertion and cheapest insertion tours are less than twice as long as an optimal tour if the triangle inequality holds (ROSENKRANTZ, STEARNS & LEWIS (1977)). It can also be shown that there exist classes of problem instances for which the length of the heuristic solution is $2 - \frac{2}{n}$ times longer than the optimal tour, thus proving that these approximation results are tight.
Recently it was shown (HURKENS (1991)) that for random or farthest insertion there exist examples where these heuristics yield tours that are 13/2 times longer than an optimal tour (although the triangle inequality is satisfied).
We have compared the nine insertion heuristics for our set of sample problems. Each heuristic was started with the cycle $(\lfloor \frac{n}{2} \rfloor, \lfloor \frac{n}{3} \rfloor, \lfloor \frac{n}{4} \rfloor)$ to get unbiased starting conditions. For the variant of cheapest insertion described in 6.2.1.6 we have set $k = 30$.
Table 6.11 displays the results (headings 1 through 9 corresponding to the insertion heuristics 6.2.1.1 through 6.2.1.9). The best solution in each row is marked with a '*'. Farthest insertion 1 performs best for our set of problems followed closely by random insertion. The fast version of cheapest insertion performs as well as the full version, the time for doing correct cheapest insertion does not pay off. In fact, the results were the same except for two cases. However, though reasonable at first sight, cheapest insertion performs significantly worse that farthest insertion. The relatively good performance of farthest insertion can be explained when observing the development of the generated tour: after few steps already, a good global outline of the final tour is obtained. Almost the same is true for random insertion. An average excess over the Held-Karp bound of 27% for the nearest insertion and of 13.5% for the farthest insertion procedure is reported in JOHNSON (1990) for random problem instances.

Problem	1	2	3	4	5	6	7	8	9
d198	13.19	3.85*	7.57	14.80	11.08	11.08	8.17	8.15	7.78
lin318	21.62	10.87	18.41	24.30	18.39	18.39	9.18*	20.02	16.27
fl417	12.50	5.48	13.37	13.37	12.39	12.39	3.29*	7.84	9.04
pcb442	20.89	13.83	16.99	29.06	21.15	21.15	12.23*	27.07	20.43
u574	22.33	11.39*	22.68	26.32	19.12	19.12	11.64	23.32	22.21
p654	10.81	6.89	11.33	5.94	5.79*	5.79*	9.87	11.30	12.64
rat783	23.04	12.09*	22.52	28.72	16.02	16.02	13.37	26.37	25.02
pr1002	18.57	10.85*	24.81	27.24	16.61	16.61	12.50	23.98	25.42
u1060	21.39	12.68	21.52	27.55	18.67	18.67	11.43*	23.94	21.58
pcb1173	25.84	14.22*	26.82	32.67	21.50	21.50	16.58	29.56	28.80
d1291	22.90	23.78	27.29	29.50	17.01*	17.01*	22.13	31.06	18.70
rl1323	31.01	18.89*	29.30	27.80	24.81	24.81	20.64	29.30	26.56
fl1400	20.28	8.45*	14.56	24.78	17.98	17.76	8.47	16.30	16.44
u1432	15.26	12.59*	20.43	20.08	12.65	12.65	12.63	23.84	20.54
fl1577	21.61	15.17*	20.04	25.21	17.08	17.08	18.70	26.66	17.97
d1655	20.18	17.09*	22.22	27.80	18.83	18.77	17.69	28.20	23.95
vm1748	21.26	13.54*	25.37	33.59	18.86	18.86	13.87	29.52	24.26
rl1889	23.82	19.10	27.74	32.70	21.24	21.24	17.30*	29.99	27.53
u2152	21.09	19.55	28.64	32.84	16.12*	16.12*	19.76	28.26	28.98
pr2392	24.70	14.32*	28.26	33.55	20.50	20.50	16.65	31.75	28.32
pcb3038	23.12	14.89*	24.54	27.84	17.08	17.08	16.69	27.57	27.28
fl3795	19.61	21.97	19.58	29.45	12.79*	12.79*	19.77	21.62	25.62
fnl4461	21.10	12.03*	27.69	28.90	15.97	15.97	12.99	28.99	28.03
rl5934	27.40	22.17	30.12	33.42	21.84*	21.84*	22.71	33.56	30.36
Average	20.98	13.99	22.16	26.56	17.23	17.22	14.51	24.51	22.24

Table 6.11 Results of insertion heuristics

Note that the quality of the solutions of the different heuristics is highly problem dependent. Running time will be addressed in the next section.

6.2.2 Fast Versions of Insertion Heuristics

As in the case of the nearest neighbor heuristic we want to give priority to edges from a candidate set to speed up the insertion heuristics.
To this end we base all our calculations on the edges contained in the candidate set. E.g., now the distance of a non-tour node v to the current partial tour is infinite if there is no candidate edge joining v to the tour, otherwise it is the length of the shortest such edge joining v to the tour. Using this principle the heuristics of the previous chapter are modified as follows.

6.2.2.1 Nearest Insertion

If there are nodes connected to the current tour by a subgraph edge then insert the node connected to the tour by the shortest edge. Otherwise insert an arbitrary node. □

6.2.2.2 Farthest Insertion 1

Among the nodes that are connected to the tour insert the one whose minimal distance to the tour is maximal. If all external nodes are not connected to the tour insert an arbitrary node. □

6.2.2.3 Farthest Insertion 2

Among the nodes that are connected to the tour insert the one whose distance to the tour is maximal. If all external nodes are not connected to the tour insert an arbitrary node. □

6.2.2.4 Farthest Insertion 3

Among the nodes that are connected to the tour insert the one whose maximal distance to the tour is minimal. If all external nodes are not connected to the tour insert an arbitrary node. □

6.2.2.5 Cheapest Insertion 1

Insert a node connected to the current tour by a subgraph edge whose insertion yields minimal additional length. If no such node exists then compute the cheapest insertion possibility. Insertion information is only updated for those nodes that are connected to the node inserted last by a subgraph edge. This way insertion information may become incorrect for some nodes since it may not be updated. □

6.2.2.6 Cheapest Insertion 2

Insert a node connected to the current tour by a subgraph edge whose insertion yields minimal additional length. If no such node exists then insert an arbitrary node. Update of insertion information is further simplified as in 6.2.1.6. □

6.2.2.7 Random Insertion

Select the node to be inserted at random where priority is given to nodes connected to the current tour by a subgraph edge. □

6.2.2.8 Largest Sum Insertion

For each node compute the sum of lengths of the subgraph edges connecting this node to the current tour. Insert the node whose sum is maximal. If all external nodes are not connected to the tour insert an arbitrary node. □

6.2.2.9 Smallest Sum Insertion

For each node compute the sum of lengths of the subgraph edges connecting this node to the current tour. Insert the node whose sum is minimal. If all external nodes are not connected to the tour insert an arbitrary node. □

We have performed the same experiment as for the heuristics in the complete graph. Results are shown in Table 6.12. Now, the advantages of farthest or random insertion are lost due to the restricted view. They still perform best but tour quality is significantly inferior than before. The cheapest insertion variants give some very bad solutions which is caused by the incomplete update of insertion information.
To visualize the CPU time for insertion heuristics we have compared five variants in Figure 6.13: Farthest insertion 6.2.1.2 ([1]), Cheapest insertion 6.2.1.5 ([2]), Cheapest insertion 6.2.1.6 ([3]), Farthest insertion 6.2.2.2 ([4]), Cheapest insertion 6.2.2.5 ([5]). The diagram shows that standard farthest insertion compares favorably with all cheapest insertion variants. Speed up using candidate graphs is considerable, but due to inferior quality there seems to be no point in using these heuristics. This will be further justified in Chapter 7.

Problem	1	2	3	4	5	6	7	8	9
d198	15.31	7.84*	9.41	13.67	13.47	13.47	10.42	13.80	11.13
lin318	25.69	20.03	18.85	23.78	42.95	42.95	18.09	17.63*	24.41
fl417	36.20	31.08	33.92	44.57	24.36*	24.36*	26.82	38.90	31.99
pcb442	28.85	18.59	20.16	19.93	29.66	29.66	20.07	14.08*	27.33
u574	22.54	17.34*	19.97	18.60	25.28	25.28	17.53	19.02	26.72
p654	48.59	46.22	36.84*	42.62	78.21	78.81	49.36	44.91	54.21
rat783	26.07	15.35*	18.31	20.00	24.90	24.90	17.47	16.11	29.58
pr1002	19.89*	21.30	29.43	20.37	26.54	26.50	22.52	20.74	28.17
u1060	25.39	17.54*	20.42	20.78	22.95	24.07	18.52	19.97	25.55
pcb1173	28.93	19.28*	21.60	21.87	34.27	34.27	21.84	22.42	28.86
d1291	31.24	25.33	26.61	27.29	20.91	20.73*	24.78	26.81	28.16
rl1323	37.34	22.46*	26.82	32.97	31.19	31.43	26.04	31.16	35.37
fl1400	30.83	31.66	28.69	29.67	85.17	94.98	19.07*	27.06	30.59
u1432	21.61	17.81	20.29	20.27	28.08	29.89	20.25	16.51*	25.52
fl1577	34.75	27.27	28.95	28.19	31.09	31.09	23.67*	29.51	36.09
d1655	28.95	23.22	23.74	26.05	33.35	35.48	22.40*	24.38	29.23
vm1748	26.05	21.07*	21.82	23.34	22.90	22.90	22.27	21.20	29.31
rl1889	35.45	25.60*	29.58	30.32	42.91	42.39	31.51	28.60	35.12
u2152	28.99	24.68	28.89	24.46	21.34*	21.34*	25.03	25.06	30.82
pr2392	27.01	23.14*	27.88	28.22	35.15	32.68	24.56	24.28	31.41
pcb3038	25.19	18.48*	21.47	19.67	25.61	25.72	20.05	20.00	28.57
fl3795	35.77	24.96*	29.32	30.18	40.31	40.62	25.80	33.85	32.41
fnl4461	23.47	16.88*	17.23	20.27	31.74	36.16	17.64	18.11	28.51
rl5934	44.63	31.26*	29.81	35.55	51.60	48.17	32.91	33.31	37.97
Average	29.53	22.85	24.58	25.94	34.33	34.91	23.28	24.48	30.29

Table 6.12 Results of fast insertion heuristics

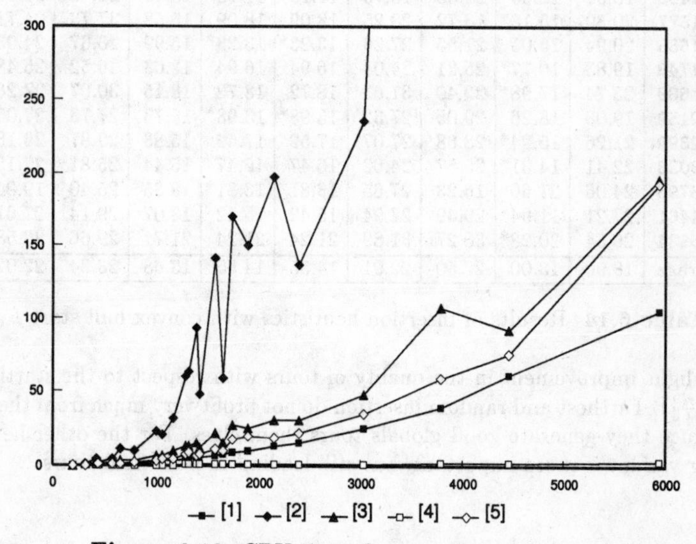

Figure 6.13 CPU times for some insertion heuristics

6.2.3 Convex Hull Start

The following observation suggests to use a specific starting tour for Euclidean problems. Let $v_1, v_2, \ldots, v_k$ be located on the boundary of the convex hull of the given points (in this order). Then, in any optimal tour, this sequence is respected (otherwise the tour would contain crossing edges and hence could not be optimal). Therefore it is reasonable to use $(v_1, v_2, \ldots, v_k)$ as starting tour for the insertion heuristics.
From the results of Chapter 4 we know that convex hulls can be computed very quickly (in time $\Theta(n \log n)$). Therefore, only negligible additional CPU time is necessary to compute this type of starting tour for the insertion heuristics in the Euclidean case. Results with the convex hull start using the standard versions of the insertion heuristics are displayed in Table 6.14.

Problem	1	2	3	4	5	6	7	8	9
d198	12.86	6.73	6.41	6.58	8.51	8.51	4.37*	7.03	7.34
lin318	15.08	10.82	19.70	16.82	11.42	11.42	7.97*	18.01	16.39
fl417	14.24	5.28	14.95	5.65	7.61	7.61	2.77 *	6.83	8.36
pcb442	16.52	10.37*	17.54	18.89	11.83	11.83	13.62	19.17	22.71
u574	17.24	9.95*	23.55	18.47	14.67	14.67	10.73	19.83	20.62
p654	17.07	3.05*	12.40	6.38	8.15	8.15	6.49	13.16	10.85
rat783	16.90	12.72	24.68	23.89	15.16	15.16	11.90*	22.31	23.80
pr1002	20.05	11.10*	25.65	21.66	14.23	14.23	13.27	26.71	21.76
u1060	22.78	10.69	24.71	22.79	16.65	16.65	10.39*	23.99	22.65
pcb1173	21.61	15.44*	26.14	26.62	19.18	19.18	18.25	28.45	26.35
d1291	25.58	21.80	25.52	26.22	14.69*	14.69*	21.03	22.06	22.37
rl1323	25.86	15.10*	28.57	25.74	20.30	20.30	20.18	27.73	27.59
fl1400	14.04	5.79*	14.62	12.05	9.73	9.73	8.35	13.69	17.90
u1432	15.34	12.65	21.06	18.73	11.73*	11.73*	13.19	22.48	21.66
fl1577	20.30	15.18*	18.72	28.25	18.09	18.09	15.58	37.73	24.71
d1655	20.94	15.05	21.55	27.26	13.23*	13.23*	15.99	26.67	24.03
vm1748	19.83	10.77*	25.31	24.04	16.94	16.94	12.03	26.52	25.48
rl1889	25.74	17.98*	29.40	31.63	18.72	18.72	18.15	30.07	27.26
u2152	19.03	18.26	29.05	27.32	13.98*	13.98*	19.73	27.73	27.05
pr2392	21.26	15.24*	28.88	27.07	17.52	17.52	15.83	29.87	26.18
pcb3038	22.41	14.31*	25.57	24.62	16.47	16.47	15.44	25.81	27.12
fl3795	24.06	21.60	16.23	27.65	13.81*	13.81*	18.35	25.40	19.04
fnl4461	22.21	11.94*	29.49	27.94	15.42	15.42	13.07	29.14	27.67
rl5934	26.54	20.28*	30.27	31.89	21.24	21.24	21.71	29.66	29.58
Average	19.90	13.00	22.50	22.01	14.55	14.55	13.68	23.34	22.02

Table 6.14 Results of insertion heuristics with convex hull start

There is a slight improvement in the quality of tours with respect to the starting tour $(\lfloor \frac{n}{2} \rfloor, \lfloor \frac{n}{3} \rfloor, \lfloor \frac{n}{4} \rfloor)$. Farthest and random insertion do not profit very much from the convex hull start since they generate good globals tours themselves. For the other heuristics, this starting variant is more important, but still leading to poor final tours.

6.2.4 Stability of Insertion Heuristics

As in the case of the nearest neighbor heuristic we also investigated how strongly our results depend on the choice of the starting tour.
To get an impression of this, we performed experiments for problems d198, lin318, and pcb442. Each heuristic was started with all possible tours consisting of just two nodes. Numbers displayed in Table 6.15 have the same meaning as in Table 6.9 for the nearest neighbor methods.

Variant	Minimum	Maximum	Average	Span	Deviation
d198					
1	8.52	17.78	12.59	9.25	1.70
2	1.53	10.72	6.08	9.18	2.11
3	4.75	16.90	7.70	12.15	2.07
4	6.95	17.00	12.14	10.05	1.77
5	7.57	15.65	10.95	8.08	1.40
6	8.25	14.18	11.01	5.93	1.30
7	2.63	8.86	5.53	6.23	1.10
8	5.87	19.99	11.55	14.13	3.56
9	3.83	13.24	7.75	9.41	1.73
lin318					
1	17.18	25.97	22.00	8.79	1.82
2	5.47	13.24	9.00	7.77	1.50
3	12.93	22.38	18.72	9.45	1.30
4	20.13	29.30	24.92	9.17	1.73
5	13.81	22.15	18.61	8.35	1.46
6	13.81	22.15	18.69	8.35	1.47
7	6.67	14.18	10.89	7.51	1.65
8	16.48	26.60	21.78	10.12	1.69
9	11.99	23.34	18.92	11.35	1.86
pcb442					
1	13.89	23.17	18.50	9.29	1.85
2	9.05	18.03	13.00	8.97	1.49
3	14.93	23.99	18.45	9.06	1.10
4	21.78	33.13	27.50	11.36	1.99
5	12.73	21.38	17.76	8.66	1.47
6	12.73	21.55	17.80	8.82	1.54
7	10.22	17.71	14.00	7.48	1.44
8	20.13	32.93	25.30	12.80	2.44
9	14.22	23.64	18.56	9.41	1.52

Table 6.15 Sensitivity analysis for insertion heuristics

Farthest insertion and random insertion also performed best here. Stability of insertion heuristics is much better than for the nearest neighbor variants. The table also shows that performance and stability are highly problem dependent.

6.3 Heuristics Using Spanning Trees

The heuristics considered so far construct in their standard versions the tours "from scratch" in the sense that they do not exploit any additional knowledge about the problem instance. In their fast variants they used the presence of a candidate subgraph to guide the tour construction.

The two heuristics to be described next use a minimum spanning tree as a basis for generating tours. They are particularly suited for problem instances obeying the triangle inequality. In this case performance guarantees are possible. Nevertheless, in principle they can also be applied to general instances.

Before describing these heuristics, consider the following observation. Suppose we are given some Eulerian tour containing all nodes of the given problem instance. If the triangle inequality is satisfied we can derive a Hamiltonian tour which is not longer than the Eulerian tour.

Let $v_{i_0}, v_{i_1}, \ldots, v_{i_k}$ be the sequence in which the nodes (including repetitions) are visited when traversing the Eulerian tour starting at v_{i_0} and returning to $v_{i_k} = v_{i_0}$. The following procedure obtains a Hamiltonian tour.

procedure obtain_tour

(1) Set $Q = \{v_{i_0}\}$, $T = \emptyset$, $v = v_{i_0}$, and $l = 1$.

(2) As long as $|Q| < n$ perform the following steps.

(2.1) If $v_{i_l} \notin Q$ then set $Q = Q \cup \{v_{i_l}\}$, $T = T \cup \{vv_{i_l}\}$, and $v = v_{i_l}$.

(2.2) Set $l = l + 1$.

(3) Set $T = T \cup \{vv_{i_0}\}$.

(4) T is a Hamiltonian tour.

end of obtain_tour

Every connection made in this procedure is either an edge of the Eulerian tour or is a shortcut replacing a subpath of the Eulerian tour by an edge connecting its two endnodes. This shortcut reduces the length of the tour if the triangle inequality is satisfied. Hence the resulting Hamiltonian tour cannot be longer than the Eulerian tour.

Both heuristics start with a minimum spanning tree and differ only in how a Eulerian graph is generated from the tree.

procedure doubletree

(1) Compute a minimum spanning tree.

(2) Double all edges of the tree to obtain a Eulerian graph.

(3) Compute a Eulerian tour in this graph.

(4) Call *obtain_tour* to get a Hamiltonian tour.

end of doubletree

Note that we get multiple edges in Step (2) which we have not allowed in our definition of graphs in Chapter 2. But it is clear that this does not create any problems. Our graph data structure is able to handle multiple edges. The running time of the algorithm is dominated by the time needed to obtain a minimum spanning tree. Therefore we have time complexity $\Theta(n^2)$ for the general TSP and $\Theta(n \log n)$ for Euclidean problems.

If we compute the minimum spanning tree with Prim's algorithm (PRIM (1957)), we could as well construct a Hamiltonian cycle along with the tree computation. We always keep a cycle on the nodes already in the tree (starting with the loop consisting of only one node) and insert the node into the current cycle which is added to the spanning tree. If this node is inserted at the best possible position this algorithm is identical to the nearest insertion heuristic. If it is inserted before or after its nearest neighbor among the cycle nodes, then we obtain the nearest addition heuristic.

CHRISTOFIDES (1976) suggested a better method to make spanning trees Eulerian. Namely, it is sufficient to add a perfect matching on the odd-degree nodes of the tree. (A **perfect matching** of a node set W, $|W| = 2k$, is a set of k edges such that each node of W is incident to exactly one of these edges.) After addition of all edges of this perfect matching, all node degrees are even and hence the graph is Eulerian.

Figure 6.16 illustrates this idea. The solid edges form a spanning tree and the broken edges form a perfect matching on the odd-degree nodes of the spanning tree.

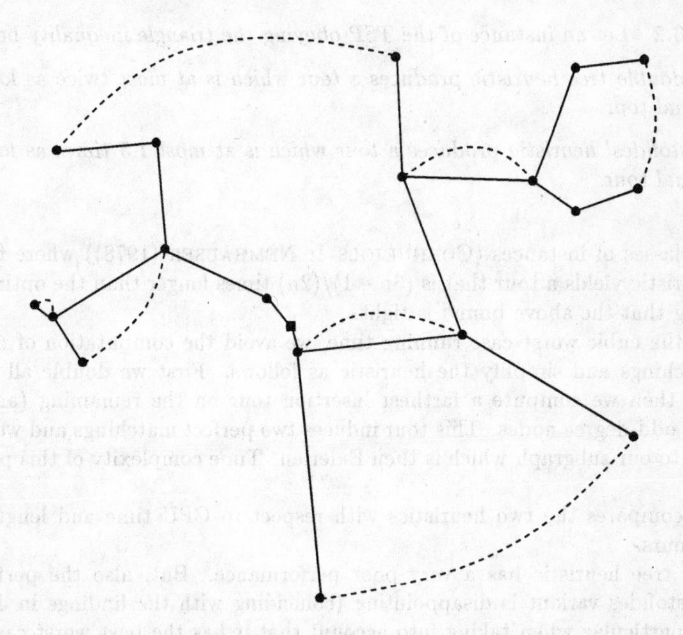

Figure 6.16 Illustration of spanning tree heuristic

The cheapest way (with respect to edge weights) to obtain a Eulerian graph is to add a minimum weight perfect matching.

procedure christofides

(1) Compute a minimum spanning tree.

(2) Compute a minimum weight perfect matching on the odd-degree nodes of the tree and add it to the tree to obtain a Eulerian graph.

(3) Compute a Eulerian tour in this graph.

(4) Call *obtain_tour* to get a Hamiltonian tour.

end of christofides

This procedure takes considerably more time than the previous one. Computation of a minimum weight perfect matching on k nodes can be performed in time $O(k^3)$ (EDMONDS (1965)). Since a spanning tree may have $O(n)$ odd-degree nodes, Christofides' heuristic has cubic worst case time.
The sequence of the edges in the Eulerian tour is not unique. So one can try to find better solutions by determining different Eulerian tours. We do not elaborate on this since the gain to be expected is small.
Since a minimum spanning tree is not longer than a shortest Hamiltonian tour and since the matching computed in Step (2) of Christofides' heuristic has weight at most half of the length of an optimal tour the following theorem holds.

Theorem 6.2 *Let an instance of the TSP obeying the triangle inequality be given.*

(i) The double tree heuristic produces a tour which is at most twice as long as an optimal tour.

(ii) Christofides' heuristic produces a tour which is at most 1.5 times as long as an optimal tour.

□

There are classes of instances (CORNUEJOLS & NEMHAUSER (1978)) where the Christofides heuristic yields a tour that is $(3n-1)/(2n)$ times longer than the optimal tour, thus proving that the above bound is tight.
Because of the cubic worst case running time, we avoid the computation of minimum weight matchings and simplify the heuristic as follows. First we double all edges to leaves, and then we compute a farthest insertion tour on the remaining (and newly introduced) odd-degree nodes. This tour induces two perfect matchings and we add the shorter one to our subgraph which is then Eulerian. Time complexity of this procedure is $O(n^2)$.
Table 6.17 compares the two heuristics with respect to CPU time and length of the generated tours.
The double tree heuristic has a very poor performance. But, also the performance of the Christofides variant is disappointing (coinciding with the findings in JOHNSON (1990)), in particular when taking into account that it has the best worst case bound among the heuristics. This is not due to our simplification, but it was observed in many experiments that it does not pay off to compute exact minimum weight perfect matchings in Step (2).

Problem	Double tree	Christofides
d198	22.62	15.67*
lin318	41.32	18.42*
fl417	36.04	24.52*
pcb442	39.64	18.59*
u574	36.29	20.08*
p654	36.89	21.73*
rat783	36.75	21.34*
pr1002	37.29	20.67*
u1060	34.30	18.97*
pcb1173	42.29	18.77*
d1291	48.16	24.31*
rl1323	39.04	14.05*
fl1400	39.40	22.10*
u1432	45.78	24.05*
fl1577	42.75	13.27*
d1655	37.47	18.92*
vm1748	31.68	21.73*
rl1889	40.50	14.00*
u2152	48.11	22.73*
pr2392	37.22	18.70*
pcb3038	43.23	20.58*
fl3795	41.38	17.25*
fnl4461	39.47	21.92*
rl5934	48.18	15.17*
Average	39.41	19.48

Table 6.17 Comparison of tree heuristics

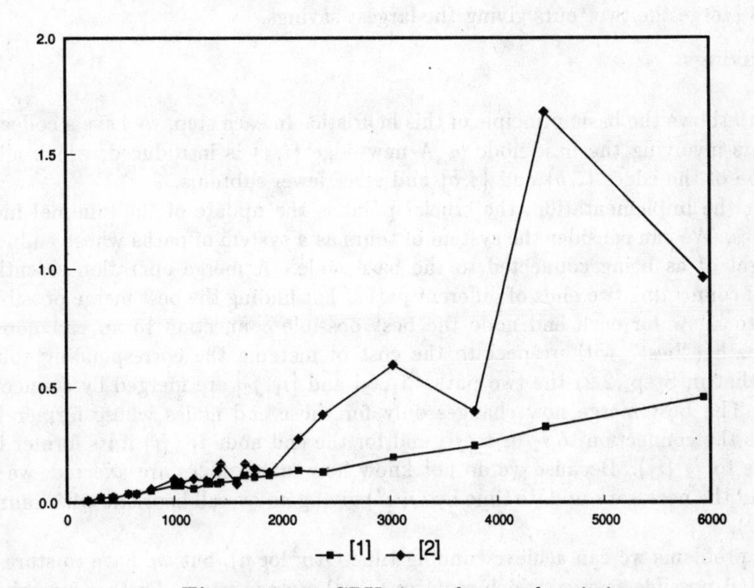

Figure 6.18 CPU times for tree heuristics

Finally we depict the CPU times for both heuristics in Figure 6.18. The time for the Christofides heuristic is highly problem dependent, because the number of odd-degree nodes in the minimum spanning tree is basically responsible for the running time.

6.4 Savings Methods and Greedy Algorithm

The savings heuristic was originally developed for vehicle routing problems (CLARKE & WRIGHT (1964)). But it can also be applied to the traveling salesman problem, since the TSP can be considered as a special vehicle routing problem involving only one vehicle. This heuristic successively merges short tours to eventually obtain a Hamiltonian tour.

procedure savings

(1) Select a base node $b \in V$ and set up the $n-1$ tours (b, v), $v \in V \setminus \{b\}$ consisting of two nodes each.

(2) As long as more than one tour is left perform the following steps.

(2.1) For every pair of tours T_1 and T_2 compute the savings that is achieved if the tours are merged by deleting in each tour an edge to the base node and connecting the two open ends. More precisely, if ub and vb are edges in different tours then these tours can be merged by eliminating ub and vb and adding edge uv resulting in a savings of $c_{ub} + c_{vb} - c_{uv}$.

(2.2) Merge the two tours giving the largest savings.

end of savings

Figure 6.19 shows the basic principle of this heuristic. In each step, we have a collection of subtours involving the base node b. A new edge $\{i, j\}$ is introduced, which allows elimination of the edges $\{i, b\}$ and $\{j, b\}$ and gives fewer subtours.
Regarding the implementation, the crucial point is the update of the minimal merge possibilities. We can consider the system of tours as a system of paths whose endnodes are thought of as being connected to the base node. A merge operation essentially consists of connecting two ends of different paths. For finding the best merge possibility we have to know for each end node the best possible connection to an end node of another path ("best" with respect to the cost of merging the corresponding tours). Suppose that in Step (2.2) the two paths $[i_1, i_2]$ and $[j_1, j_2]$ are merged by connecting i_2 to j_1. The best merge now changes only for those end nodes whose former best merge was the connection to i_2 or to j_1, and for the end node i_1 (j_1) if its former best merge was to j_2 (i_1). Because we do not know how many nodes are affected, we can only bound the necessary update time by $O(n^2)$ giving an overall heuristic with running time $O(n^3)$.
For small problems we can achieve running time $O(n^2 \log n)$, but we have to store the matrix of all possible savings which requires $O(n^2)$ storage space. Further remarks on the Clarke/Wright algorithm can be found in POTVIN AND ROUSSEAU (1990).

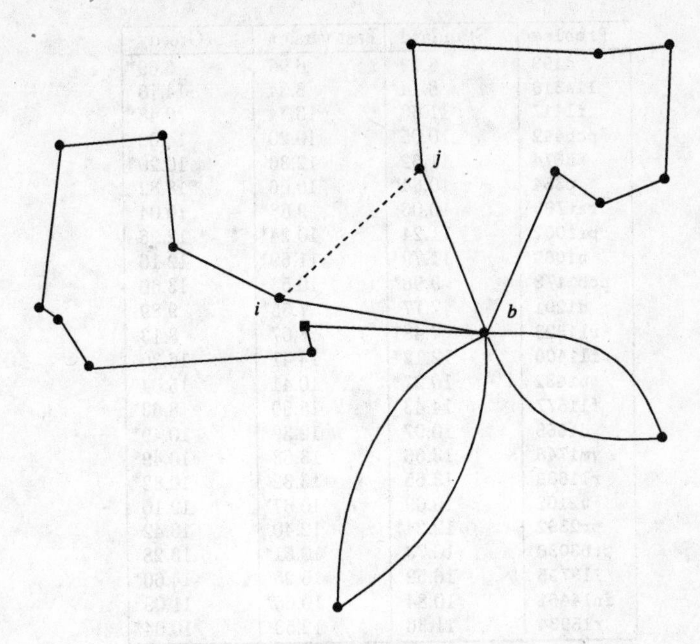

Figure 6.19 Illustration of savings heuristic

We implemented a fast version of this heuristic making use of a candidate subgraph. Now, merge operations are preferred that use a candidate edge for connecting two paths. The update is simplified in that for a node whose best merge possibility changes only candidate edges incident to that node are considered for connections. If during the algorithm an endnode of a path becomes isolated, since none of its incident subgraph edges is feasible anymore, we compute its best merge possibility by enumeration.

Instead of using the particular savings criterion given above, we could as well build systems of paths of increasing length in a different way. E.g., we obtain another heuristic, if we discard the base node, start with a system of n paths of length 0, and then check in each step if the shortest edge not considered so far can be used to join two paths. This is exactly the greedy approach of section 2.5 applied to the TSP. Let $E_n = \{e_1, e_2, \ldots, e_m\}$ (where $m = n(n-1)/2$) be the edges of K_n.

procedure TSP_greedy

(1) Sort E_n such that $c_1 \leq c_2 \leq \ldots, \leq c_m$.

(2) Set $T = \emptyset$.

(3) For $i = 1, 2, \ldots, m$:

(3.1) If $T \cup \{e_i\}$ can be extended to a Hamiltonian tour (or is a Hamiltonian tour), then set $T = T \cup \{e_i\}$.

end of TSP_greedy

Problem	Standard	Fast version	Greedy
d198	8.72	6.96	6.69*
lin318	8.14*	8.34	14.13
fl417	11.63	13.74	9.48*
pcb442	10.05*	10.20	14.62
u574	11.62	12.36	10.20*
p654	10.64*	10.66	18.82
rat783	10.06	9.88*	10.04
pr1002	11.24	10.24*	12.96
u1060	11.79	11.69*	12.16
pcb1173	9.98*	10.53	13.66
d1291	7.77	7.55*	9.89
rl1323	7.48*	8.07	8.13
fl1400	12.12*	14.41	16.30
u1432	10.27*	10.41	15.04
fl1577	14.43	15.90	8.63*
d1655	10.97	12.39	10.49*
vm1748	13.66	13.68	10.49*
rl1889	13.65	13.32	10.82*
u2152	10.69	10.67*	12.10
pr2392	12.24*	12.40	13.42
pcb3038	10.73	10.51*	13.28
fl3795	16.52	16.25	14.60*
fnl4461	10.84	10.65*	11.09
rl5934	11.36	12.53	10.04*
Average	11.11	11.39	11.96

Table 6.20 Results for savings heuristics and greedy

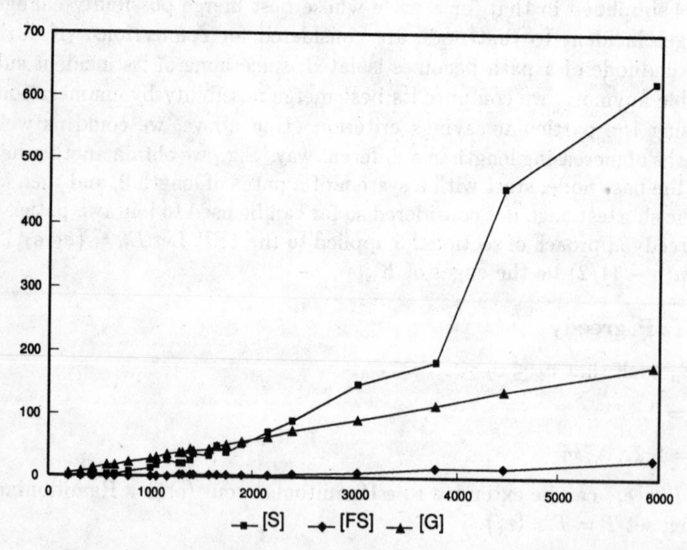

Figure 6.21 CPU times for savings heuristics and greedy

For TSPs with triangle inequality, the greedy tour can be almost $\log n$ times as long as an optimal tour (for details see FRIEZE (1979)).
The check in Step (3.1) can be accomplished in constant type with appropriate data structures. Due to the sorting step, the running time is $\Omega(n^2 \log n)$ and $\Omega(n^2)$ space is needed. We have therefore not implemented the pure greedy algorithm, but have "approximated" it for test purposes as follows. We first compute the 100 nearest neighbor subgraph and apply the greedy algorithm using only this edge set. This will usually result in a system of paths which we then connect using the savings heuristic with some randomly chosen base node.
We compared these three approaches with respect to obtained quality and running time. The results are shown in Table 6.20 and the CPU times in Figure 6.21 (diagrams [S], [FS], and [G] corresponding to the running times of the standard savings, the fast savings, and the approximate greedy heuristic, respectively). We used $\lfloor \frac{n}{2} \rfloor$ as the base node for each problem. For the fast version we used the 10 nearest neighbor subgraph. Surprisingly, the simplified heuristic yields solutions of the same quality at a considerably reduced running time.
We have also conducted an experiment concerning the stability of this heuristic. Table 6.22 displays for some problems the results that were obtained when applying the fast savings heuristic for every possible base node.

	Minimum	Maximum	Average	Span	Deviation
d198	2.97	14.26	6.54	11.29	2.25
lin318	4.99	13.65	8.46	8.66	1.53
pcb442	6.52	13.56	9.75	7.04	1.29
u1060	9.17	14.28	11.44	5.11	0.82

Table 6.22 Sensitivity analysis for savings heuristic

Comparing these results with the figures given in Tables 6.9 and 6.15 we see that the savings heuristic gives much better results and is more stable than nearest neighbor or insertion heuristics.

6.5 Comparison of Construction Heuristics

We close this chapter with a comparative assessment of all construction heuristics discussed in this chapter.
Absolute qualities of the heuristics have been given in previous tables. We now assess the performance of the construction heuristics in a different way. Namely, we compare the tour generated by a heuristic not with the best known lower bound for the respective problem instance, but with the best solution found by any of the other heuristics. Qualities are computed with respect to these best solutions and are listed in Table 6.23. This way we classify each heuristic relative to the other construction heuristics. In addition we give the number of best solutions found by every heuristic.
The clear winners with respect to absolute quality as well as with respect to relative quality are the savings heuristics, and because of the considerably less running time we declare the fast implementation of the savings heuristics to be the best construction

heuristic. For this heuristic we can expect an average quality of about 11 – 12%. Moreover due to Table 6.23 we can expect that, on the average, no other construction heuristic can produce a solution that is more than 1 – 2% better than the savings solution.

Heuristic	No. of best solutions	Relative quality
Savings (standard)	5	1.96
Savings (fast)	6	2.22
Savings (approx. greedy)	5	2.75
Farthest insertion 1 (convhull)	3	3.64
Random insertion (convhull)	3	4.26
Farthest insertion 1	1	4.55
Random insertion	–	5.03
Cheapest insertion (convhull)	–	5.08
Fast cheapest insertion (convhull)	–	5.08
Fast cheapest insertion	–	7.53
Cheapest insertion	1	7.54
Christofides	–	9.67
Nearest insertion (convhull)	–	9.99
Nearest insertion	–	10.98
Nearest neigbor variant 4	–	11.55
Farthest insertion 3 (convhull)	–	11.83
Minimum sum insertion (convhull)	–	11.88
Farthest insertion 2	–	12.02
Minimum sum insertion	–	12.07
Farthest insertion 2 (convhull)	–	12.33
Farthest insertion (fast)	–	12.81
Maximum sum insertion (convhull)	–	13.07
Random insertion (fast)	–	13.18
Maximum sum insertion	–	14.14
Maximum sum insertion (fast)	–	14.31
Farthest insertion 2 (fast)	–	14.37
Farthest insertion 3 (fast)	–	15.66
Farthest insertion 3	–	16.03
Nearest neigbor variant 2	–	16.20
Nearest neigbor variant 3	–	16.61
Standard nearest neigbor	–	16.83
Nearest neigbor variant 5	–	17.77
Nearest insertion (fast)	–	18.93
Minimum sum insertion (fast)	–	19.60
Cheapest insertion (fast)	–	23.39
Fast cheapest insertion (fast)	–	23.94
Double tree	–	27.90

Table 6.23 Comparison of construction heuristics

Finally, we give a diagram of the CPU time of the fast savings heuristic for problem instances up to 18512 nodes. To show that the preprocessing times for computing the Delaunay graph and the 10 nearest neighbor graph are small we give the CPU times with (diagram [2]) and without (diagram [1]) preprocessing time. Figure 6.24 shows that preprocessing time is negligible.

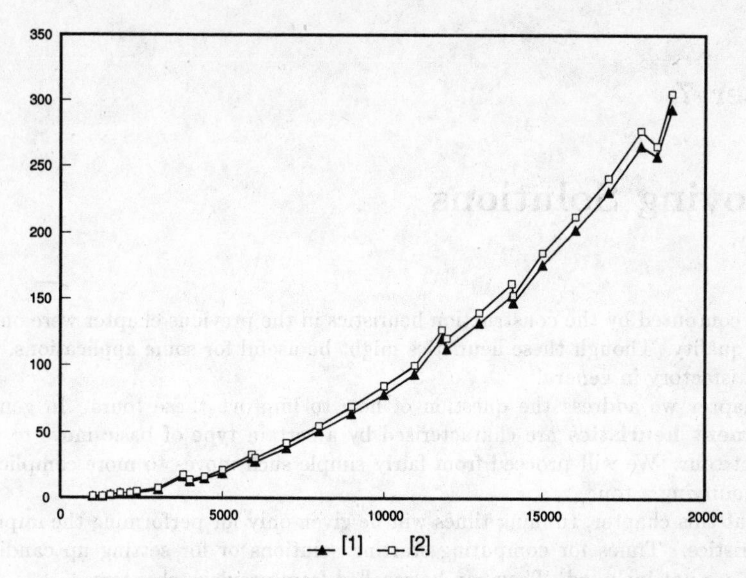

Figure 6.24 CPU time for fast savings heuristic

Increasing effort and CPU time with respect to the fast savings heuristic do not pay off. If one has to employ substantially faster heuristics then one should either use the Christofides heuristic or the variant of the nearest neighbor heuristic where forgotten nodes are inserted.

If still less CPU time is required one has to be satisfied with the results provided by the double tree heuristic. All other heuristics do not seem to have particular advantages. So, if one considers pure construction of tours they can be safely disregarded. The situation is slightly different when using improvement heuristics to modify tours. This will be addressed in the next chapter.

Chapter 7

Improving Solutions

The tours computed by the construction heuristics in the previous chapter were only of moderate quality. Though these heuristics might be useful for some applications, they are not satisfactory in general.

In this chapter we address the question of how to improve these tours. In general, **improvement heuristics** are characterized by a certain type of basic move to alter the current tour. We will proceed from fairly simple such moves to more complicated ways of modifying a tour.

Throughout this chapter, running times will be given only for performing the improvement heuristics. Times for computing starting solutions or for setting up candidate subgraphs are not included. They can be recalled from previous chapters.

7.1 Node and Edge Insertion

It is clear that we have to restrict ourselves to fairly simple modifications of the current tour if we do not want to spend too much CPU time.

We start with two straightforward modifications of the current tour which do not involve much work for updating the tours after having performed these modifications.

The first alteration of the tour is called **node insertion**. It consists of removing a node from the current tour and reinserting it at the best possible location. This operation is visualized in the upper part of Figure 7.1. Having found an improving node insertion move, the tour can be updated in constant time. It takes time $O(n^2)$ to check if tour improvement by node insertion is possible since we have to examine for every node every possible insertion point.

The heuristic based on this operation is the following.

procedure node_insertion

(1) Let T be the current tour.

(2) Perform the following until failure is obtained.

(2.1) For every node $i = 1, 2, \ldots, n$:
Examine all possibilities to insert i at a different position in the tour. If it is possible to decrease the tour length this way, then choose the such best node insertion move and update T.

(2.2) If no improving move could be found, then declare failure.

end of node_insertion

Running time of the procedure node insertion cannot be polynomially bounded in the worst case since it depends on the reductions achieved by the tour modifications. In the worst case we may have the situation that every improvement decreases the tour length by only one unit and hence the running time depends on the tour length (and not on the logarithm of the length).

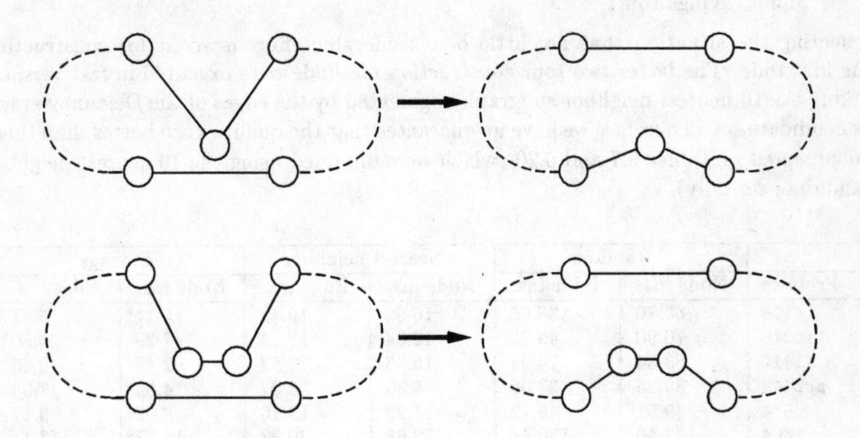

Figure 7.1 Node and edge insertion moves

A similar operation is **edge insertion**. Here, instead of a single node an edge is removed from the tour and reinserted at the best possible position. The operation is depicted in the lower part of Figure 7.1. Note that at every insertion point there are two possibilities for connecting the eliminated edge (obtained by switching its endnodes).
The corresponding heuristic is implemented in the same way as the node insertion heuristic.

procedure edge_insertion

(1) Let T be the current tour.

(2) Perform the following until failure is obtained.

(2.1) For every node $i = 1, 2, \ldots, n$:
Examine all possibilities to insert the edge between i and its successor at a different position in the tour. If it is possible to decrease the tour length this way choose the best such edge insertion move and update T.

(2.2) If no improving move could be found, then declare failure.

end of edge_insertion

Concerning the running time the same remarks as for node insertion apply. In particular, it also takes time $O(n^2)$ to check if an improving edge insertion move exists at all because for every tour edge every possible insertion point has to be checked.

To get an impression of the usefulness of these two heuristics we applied them on three types of starting tours:

1) a random tour,
2) a nearest neighbor tour,
3) and a savings tour,

reflecting the situation that no, little or considerable effort is spent for constructing the first tour. The latter two tour construction methods were executed in fast versions taking the 10 nearest neighbor subgraph augmented by the edges of the Delaunay graph as candidate set. Therefore we have no guarantee that the qualities are better than those documented in Tables 6.7 and 6.20 (which were obtained using the 10 nearest neighbor candidate set only).

	Random		Nearest neighbor		Savings	
Problem	Node ins.	Edge ins.	Node ins.	Edge ins.	Node ins.	Edge ins.
d198	65.70	38.05	16.98	19.49	5.11*	5.87
lin318	70.90	43.23	12.34	13.14	7.29	6.50*
fl417	83.34	74.99	19.96	22.53	2.47*	5.86
pcb442	30.25	33.09	8.30	8.37	4.54*	7.93
u574	39.51	45.72	14.23	12.15	8.02*	9.17
p654	177.40	130.44	32.68	30.33	11.77*	13.11
rat783	51.18	53.46	12.97	15.31	6.36*	8.09
pr1002	89.53	65.00	13.62	15.80	6.93*	8.49
u1060	76.39	60.49	17.59	17.17	6.66*	7.97
pcb1173	54.19	52.66	15.00	14.76	5.34*	8.38
d1291	71.19	73.08	16.24	18.03	5.07*	6.38
rl1323	89.13	78.66	15.00	18.17	5.26*	7.27
fl1400	232.27	140.03	24.77	23.58	12.44	9.36*
u1432	63.11	53.58	13.97	14.80	5.87*	7.97
fl1577	187.61	135.10	21.20*	21.29	21.61	22.32
d1655	84.74	77.50	11.42	11.80	9.30*	11.82
vm1748	73.22	77.48	13.90	14.93	7.75*	9.37
rl1889	86.22	80.93	16.38	17.78	7.19*	8.25
u2152	74.82	90.18	15.54	16.48	7.24*	9.53
pr2392	88.33	81.35	13.25	15.26	7.53*	9.54
pcb3038	56.39	70.74	13.08	14.20	6.10*	7.92
fl3795	318.65	202.75	29.49	29.34	21.12*	22.63
fnl4461	72.73	81.14	12.27	14.34	6.65*	8.75
rl5934	95.56	104.37	17.89	19.08	9.20*	11.30
Average	97.18	81.00	16.59	17.42	8.20	9.74

Table 7.2 Results of node and edge insertion heuristics

Table 7.2 displays the results (best solutions for each problem are marked by a '*'). The performance of these simple heuristics turns out to be rather poor, exhibiting their limited ability to modify a tour. In particular, with a random start, the heuristics terminate with solutions far away from an optimal solution. On the average, we obtain tours twice as long as a shortest tour. Nearest neighbor tours can be shortened by about 10% and savings tours by only 2–3%.

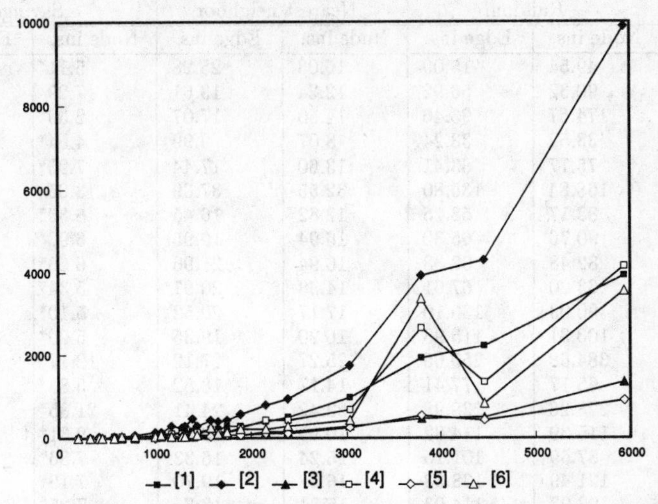

Figure 7.3 CPU times for node and edge insertion

We give the CPU times in Figure 7.3, where diagrams [1] through [6] correspond to the six columns of Table 7.2. E.g., it took 165 minutes to perform the edge insertion heuristic on a random tour for problem `rl5934`. Starting with a reasonable tour reduces CPU time drastically, since search for an improving move terminates earlier.

But, since implementation of node or edge insertion is not difficult (because of the easy tour update) it is worthwhile to think about speedup possibilities.

To reduce the CPU time one has to limit the number of moves that are checked. To this end we make use of a candidate subgraph. In Step (2.1) of the heuristics, only those moves are considered now that introduce a candidate edge into the tour. In the experiments that are documented in Table 7.4 we used the 10 nearest neighbor subgraph augmented by the edges of the Delaunay graph as the candidate set.

The quality of the tours found by this restricted search has now decreased. There is a substantial loss in quality for random starts, but only a slight decrease for the savings start. In particular, the moderate performance for random starting tours shows that the limitation of the possible moves by the candidate set is restrictive in this case. Note that bad tours can still be delivered after having performed this simple improvement. For example, we can obtain tours that are four or even five times longer than an optimal tour.

By using candidate sets, CPU time is decreased by a large factor. For example, it now takes 16 seconds to perform the limited edge insertion on a random tour for problem `rl5934`. Our restriction allows more node insertion than edge insertion moves, therefore now the CPU times for the node insertion versions are larger than for the respective edge insertions. CPU times are depicted in Figure 7.5.

To visualize that these simple heuristics usually give unsatisfactory tours we show in Figure 7.6 a tour that was obtained from a random tour for problem `rd100` by performing node as well as edge insertions as long as improvements were possible. This tour is 16.5%

	Random		Nearest neighbor		Savings	
Problem	Node ins.	Edge ins.	Node ins.	Edge ins.	Node ins.	Edge ins.
d198	49.54	15.09	16.03	25.28	5.11*	5.84
lin318	90.32	56.92	12.24	13.61	7.29	6.97*
fl417	174.67	95.46	14.30	17.07	2.39*	5.98
pcb442	33.52	33.24	8.07	9.99	4.54*	7.82
u574	75.77	63.41	13.60	17.44	7.90*	9.31
p654	168.84	136.80	32.55	37.69	13.05	12.64*
rat783	83.17	58.15	12.82	16.45	6.36*	8.02
pr1002	90.76	65.30	16.94	19.95	6.92*	8.50
u1060	82.48	88.38	16.94	21.96	6.66*	7.91
pcb1173	63.20	67.61	14.38	20.97	5.34*	8.41
d1291	90.03	130.10	17.17	20.52	5.10*	6.85
rl1323	103.31	115.98	16.20	19.35	5.29*	7.28
fl1400	384.68	250.66	25.27	27.12	12.17*	12.87
u1432	65.17	77.41	14.17	18.52	5.83*	7.75
fl1577	228.26	238.98	22.42	24.41	21.85*	22.82
d1655	115.39	114.82	10.99	13.54	9.31*	11.94
vm1748	87.59	101.15	15.24	16.33	7.86*	9.68
rl1889	121.49	138.38	16.38	19.70	7.19*	8.68
u2152	93.97	114.03	15.81	18.34	7.25*	9.46
pr2392	97.34	91.24	13.94	17.23	7.53*	9.68
pcb3038	75.01	91.15	13.11	16.98	6.12*	8.04
fl3795	421.10	321.70	30.22	34.29	22.07*	26.11
fnl4461	86.28	115.48	12.47	16.52	6.65*	8.75
rl5934	190.74	207.06	19.87	22.24	9.21*	11.77
Average	128.03	116.19	16.71	20.23	8.29	10.13

Table 7.4 Results of fast node and edge insertion heuristics

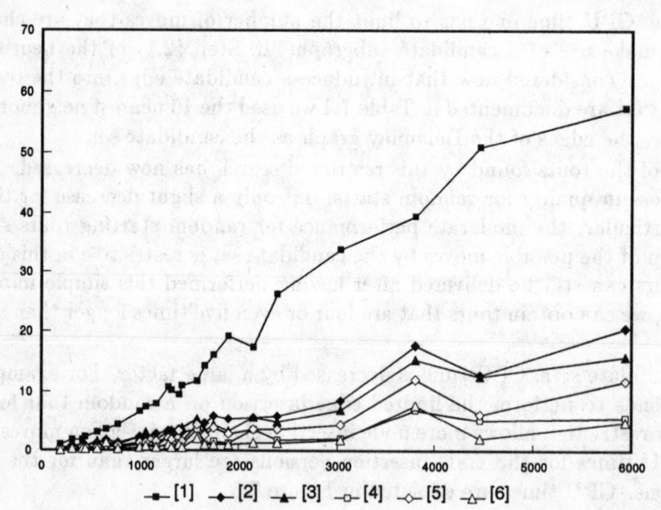

Figure 7.5 CPU times for fast node and edge insertion

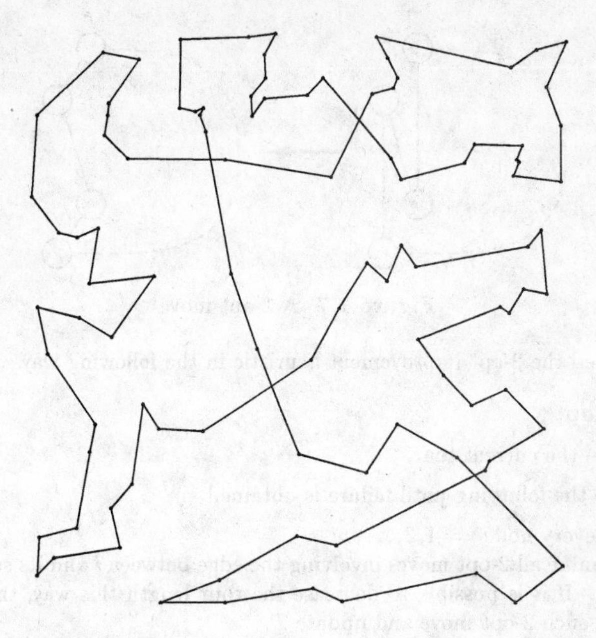

Figure 7.6 A tour found by node and edge insertion

above optimum and verifies that simple moves are not at all sufficient to turn a random tour into an acceptable tour.
Furthermore, it can be observed that these heuristics very slowly approach their local optimum. It is usually necessary to check a large number of useless moves before an improving move is found. These simple insertion methods (in particular node insertion) are suited, however, for finding further slight improvements of reasonable tours.
There are additional techniques to speed up the search for improving moves. We will consider these techniques in the following sections. Their application to these simple heuristics does not seem to be worthwhile.

7.2 2-Opt Exchange

The next tour improvement approach is motivated by the following observation. Consider the Euclidean case. If a tour crosses itself it can be easily shortened. Namely, erase two edges that cross and reconnect the resulting two paths by edges that do not cross (this is always possible). The new tour is shorter than the old one.
A **2-opt move** in general consists of eliminating two edges and reconnecting the two resulting paths in a different way to obtain a new tour. Note that there is only one way to reconnect the paths (if we do not use the eliminated edges again). For improving a tour this way, the two eliminated edges do not necessarily have to cross (even in the Euclidean case). Figure 7.7 displays a 2-opt move involving non-crossing edges.

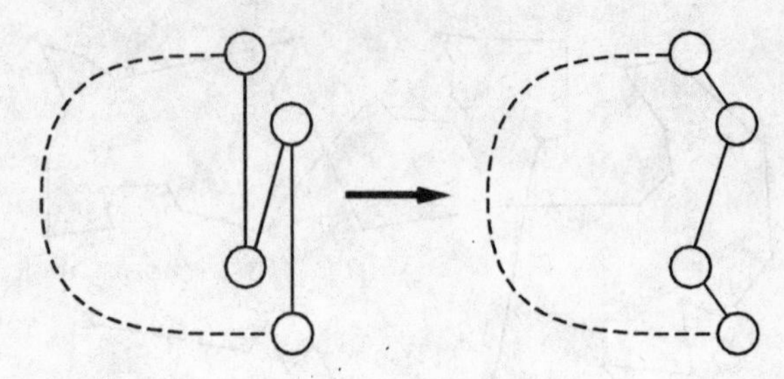

Figure 7.7 A 2-opt move

We implemented the 2-opt improvement heuristic in the following way.

procedure 2-opt

(1) Let T be the current tour.

(2) Perform the following until failure is obtained.

(2.1) For every node $i = 1, 2, \ldots, n$:
Examine all 2-opt moves involving the edge between i and its successor in the tour. If it is possible to decrease the tour length this way, then choose the best such 2-opt move and update T.

(2.2) If no improving move could be found, then declare failure.

end of 2-opt

Again, since we have assumed integral data and since we perform only improving moves, the procedure runs in finite time. But, in the worst case, it can only be guaranteed that an improving move decreases the tour length by at least one unit. No polynomial worst case bound for the number of moves to reach a local minimum can be given. Checking whether an improving 2-opt move exists takes time $O(n^2)$ because we have to consider all pairs of tour edges.
We performed this 2-opt heuristic starting with the same three tours as in the case of node and edge insertion, namely

1) a randomly generated tour,
2) a nearest neighbor tour,
3) and a tour found by the savings heuristic.

Table 7.8 shows the results. The 2-opt procedure performs much better than the insertion heuristics. In particular, random starting tours can now, on the average, be turned into tours of about 15% above the optimum. An interesting observation is that performance with nearest neighbor or savings starts is similar. Qualities differ not much, yielding tours of about 9% above optimality on average. For a particular problem instance, however, results can be rather different and none of these two starting tours supersedes the other one.
For random problem instances, JOHNSON (1990) and BENTLEY (1992) report excess of 2-optimal tours of 6.5%, resp. 8.7%, over an approximation of the Held-Karp lower bound.

Problem	Random	Nearest N.	Savings
d198	8.04	3.18*	5.29
lin318	13.05	5.94*	8.43
fl417	12.25	7.25	5.38*
pcb442	12.64	7.82	7.70*
u574	14.24	7.02*	8.82
p654	12.40	12.37	8.66*
rat783	12.31	8.39	8.03*
pr1002	14.91	8.48*	9.07
u1060	13.05	9.11	8.94*
pcb1173	12.85	9.42	7.78*
d1291	17.72	9.62	6.22*
rl1323	15.89	7.88	6.56*
fl1400	12.50	9.79	8.85*
u1432	14.24	10.07	8.83*
fl1577	21.42	8.15*	12.59
d1655	16.42	8.29*	12.36
vm1748	12.74	8.58*	9.20
rl1889	14.22	8.64	8.55*
u2152	19.89	10.02	9.64*
pr2392	16.20	8.27*	9.57
pcb3038	16.29	8.34*	8.36
fl3795	13.52	8.57*	11.37
fnl4461	14.09	7.77*	8.90
rl5934	21.07	9.19*	10.98
Average	14.67	8.42	8.75

Table 7.8 Results of standard 2-opt for different starting tours

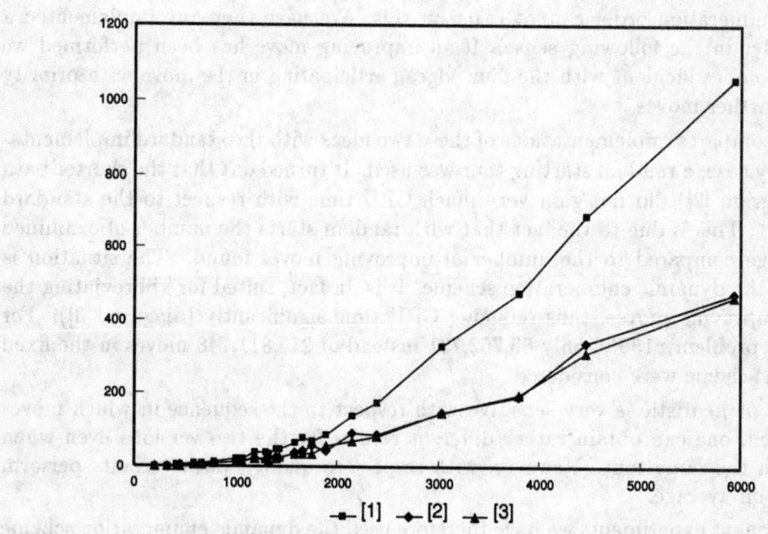

Figure 7.9 CPU times for standard 2-opt heuristic

The relatively good results for the nearest neighbor start are due to the fact that nearest neighbor tours are locally quite reasonable. The few very "bad" edges are easily eliminated using the 2-opt heuristic. The random start exhibits that also 2-opt moves are not powerful enough to give acceptable tours when started without additional knowledge (as it is provided by the other heuristics). CPU times are displayed in Figure 7.9, diagrams [1] to [3] corresponding to the different choices 1) to 3) of starting tours. Running times are considerable, about equal for the nearest neighbor and savings starts, and substantially higher for the random start. This is due to the fact that more moves are examined and performed to turn a random tour into a locally optimal tour.

We discuss some ideas for improving the running time of the 2-opt heuristic.

Recall section 2.4.5 explaining the necessity of having an imposed direction on the tour. Having performed a move, the direction has to be updated for one of the two segments of the tour. One could save CPU time if the direction on the longer path is maintained and if only the shorter path is reversed. Therefore, a first reasonable improvement is to incorporate this shorter path update. It can be accomplished by using an additional array giving the rank of the nodes in the current tour (an arbitrary node receives rank 1, its successor gets rank 2, etc.). Having initialized these ranks we can determine in constant time which of the two parts is shorter, and the ranks have to be updated only for the nodes in the shorter part.

A second idea is to avoid the unguided search for improving moves. A standard approach is to use a fixed enumeration scheme, e.g., always scanning the nodes in Step (2.1) of the heuristic in the sequence $1, 2, \ldots, n$ and checking if a move containing the edge from node i to its successor in the current tour can participate in an allowed move (taking restrictions based on the candidate set into account). But usually, one observes that in the neighborhood of a successful 2-opt move more improving moves can be found. The fixed enumeration order cannot react on this. We have therefore implemented a dynamic order in the following sense. If an improving move has been performed we consider the nodes incident with the four edges participating in the move with priority for finding further moves.

Figure 7.10 compares implementations of these two ideas with the standard implementation. In every case, a random starting tour was used. It turned out that the shorter path update (diagram [2]) did not gain very much CPU time with respect to the standard (diagram [1]). This is due to the fact that with random starts the number of examined moves is large compared to the number of improving moves found. The situation is different for the dynamic enumeration scheme. It is, in fact, suited for abbreviating the search for improving moves, thus reducing CPU time significantly (diagram [3]). For example, for problem `rl5934` only 85,762,731 instead of 215,811,748 moves in the fixed enumeration scheme were considered.

Since the 2-opt heuristic is very sensitive with respect to the sequence in which moves are performed, one can obtain rather different results for the two versions even when starting with the same tour. However, with respect to quality, both variants perform equally well on average.

In the subsequent experiments we have therefore used the dynamic enumeration scheme as well as the shorter path update (which is more effective if candidate subgraphs are employed).

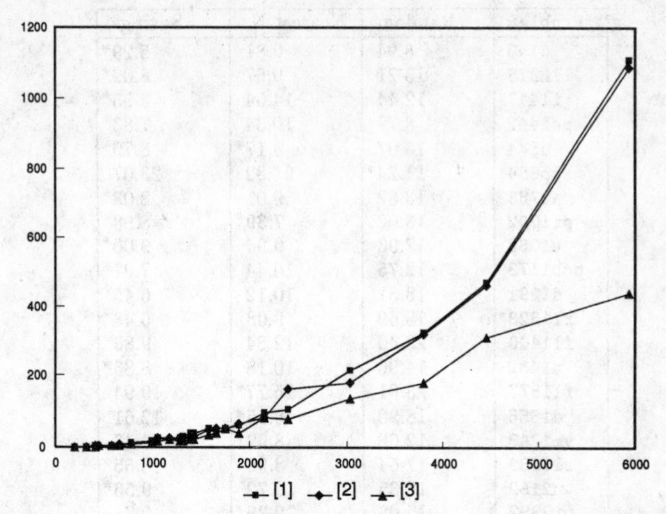

Figure 7.10 Influence of shorter path update and dynamic enumeration

Still an enormous amount of CPU time is needed in this implementation. This is due to the fact that most of the time is spent in examining useless 2-opt moves. The only possibility to reduce the running time is to restrict the number of 2-opt moves that are checked. Here again, a candidate subgraph comes into play. We compute a reasonable candidate subgraph and reduce the number of 2-opt moves by considering only moves that introduce at least one candidate edge into the current tour. Of course, we do this again at the risk of missing possible improvements.

In the following experiment, we used the same starting tours as before, but this time we used the 10 nearest neighbor subgraph augmented by the Delaunay graph to reduce the number of steps of the 2-opt heuristic. The chosen candidate subgraph is also applied for speeding up the nearest neighbor and the savings heuristic.

Table 7.11 shows the length of the tours obtained and Figure 7.12 displays the CPU time (as usual not including the candidate set computations).

Quality is slightly worse than before. For particular problems, as e.g., problems **p654** or `fl1577`, however, limited search can lead to substantial longer tours. On average, for the reasonable starting tours, quality is reduced from about 9% to about 10%.

Before we decided to choose this candidate subgraph we have also conducted experiments with the pure 10 nearest neighbor subgraph. In this case we obtained some very bad results with the following explanation. There are some problem instances consisting of clusters of points which are quite far apart from each other. Since the 10 nearest neighbor subgraph contains no edges connecting these clusters, the restricted heuristic often fails to recover from bad configurations. Several long edges cannot be eliminated. To come up with better solutions, we have to provide connections between the clusters. Therefore we added the edges of the Delaunay graph to the candidate set.

CPU times could be drastically reduced compared to the complete enumeration of all possible moves.

Problem	Random	Nearest N.	Savings
d198	8.94	6.81	5.29*
lin318	13.21	9.67	8.00*
fl417	12.44	14.54	5.55*
pcb442	8.79	10.34	7.83*
u574	15.07	8.17*	8.79
p654	11.24*	21.82	23.07
rat783	13.82	9.01	8.03*
pr1002	13.02	7.39*	8.98
u1060	17.93	9.94	9.00*
pcb1173	12.75	10.44	7.97*
d1291	18.81	10.12	6.45*
rl1323	18.69	9.08	6.48*
fl1400	23.40	12.84	9.86*
u1432	14.96	10.18	8.83*
fl1577	23.91	15.77*	19.94
d1655	18.90	8.25*	12.61
vm1748	12.68	8.92*	9.43
rl1889	17.54	8.98	8.55*
u2152	18.65	10.72	9.56*
pr2392	15.08	9.29*	9.74
pcb3038	13.96	8.49	8.47*
fl3795	52.35	11.69*	18.85
fnl4461	12.64	7.70*	8.95
rl5934	19.65	11.17	10.88*
Average	17.02	10.47	10.05

Table 7.11 Results of fast 2-opt for different starting tours

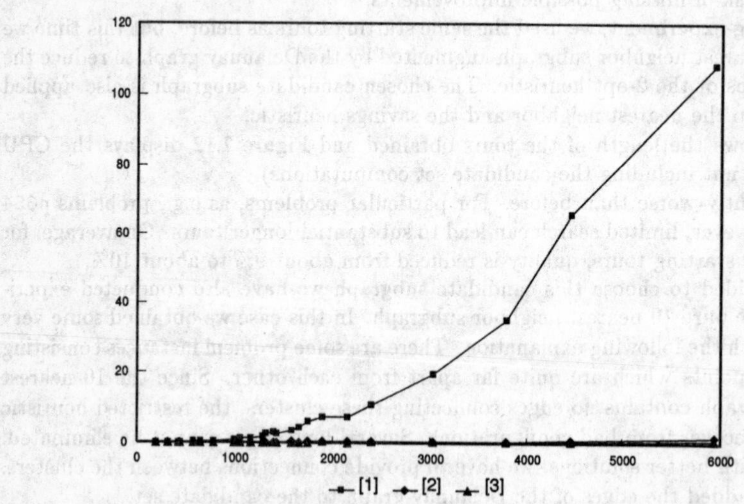

Figure 7.12 CPU times for fast 2-opt heuristic

The comparison shows that random starts are not advisable for the 2-opt heuristic and that restricted search to speed up the computations is necessary. Because of the limited power of 2-opt, it is recommended to start out with reasonable tours.
We will briefly address a further idea that can be exploited to speed up the 2-opt exchange heuristic. In section 2.4 we have discussed how a suitable data structure can be employed to speed up tour updates arising from 2-opt moves. We can use this data structure to avoid updating the direction of one part of the tour after each 2-opt move. Rather, we collect a sequence of improving 2-opt moves before updating the tour. Recall that it takes expected $O(n)$ time to update a tour according to a 2-opt move (even if only the direction on the shorter of the two paths is altered).
In experiments we used the interval data structure to store up to 50 2-opt moves before really updating the tour. It turned out, that (at least in our implementation) signifcant reduction of CPU time was only observed for very large problems, with sizes above 10,000 nodes, say.
In our further 2-opt experiments we have therefore not employed this idea. The interval technique, however, turned out to be very useful for speeding up the Lin-Kernighan heuristic to be discussed in section 7.5.
We have run some experiments to get an impression about the stability of the 2-opt heuristic. For each of the six selected problems we used a random starting tour in the first experiment and a nearest neighbor starting tour in the second one. The candidate subgraph consisted of the 10 nearest neighbor subgraph augmented by the Delaunay graph. We applied the dynamic enumeration strategy and had a run for each of the starting lists $L_i = (i, i+1, \ldots, n, 1, 2, \ldots, i-1)$. Results are documented in Table 7.13.

Variant	Minimum	Maximum	Average	Span	Deviation
rd100					
1	3.20	21.01	10.84	17.81	3.21
2	4.74	14.25	9.90	9.51	1.82
d198					
1	2.38	19.03	9.91	16.65	2.92
2	5.86	8.64	6.48	2.78	0.54
lin318					
1	5.85	21.70	11.67	15.85	2.56
2	7.92	10.97	9.37	3.05	0.87
fl417					
1	8.49	122.66	47.38	114.17	17.47
2	3.58	7.74	5.70	4.16	1.18
pcb442					
1	8.19	18.90	13.05	10.71	1.91
2	5.18	9.69	6.43	4.51	0.76
u574					
1	7.58	21.67	13.16	14.09	2.04
2	5.77	9.63	7.35	3.86	1.06

Table 7.13 Stability analysis of the 2-opt heuristic

Table 7.13 shows that there can be substantial differences if the nodes are considered in a different sequence. In particular, for the random start, very long tours can be obtained. With nearest neighbor starts we obtain a reasonable average performance with little

deviation. This suggests to run several 2-opt improvements on a particular problem. Due to the small deviation, one can then expect to achieve average performance.
The results displayed so far and further experiments that are not documented here show that the 2-opt exchange heuristic outperforms the node and edge insertion heuristics by far. They cannot compete at all. On the other hand, node insertion can help to improve tours and only a few special node insertion moves are also 2-opt moves. Since node insertion is not difficult to implement, we considered in a final experiment a combination of 2-opt and node insertion.
In addition, though not of practical relevance, we first performed the combined heuristic enumerating all possible moves.

Problem	Random	Nearest N.	Savings
d198	4.39	2.26*	4.25
lin318	10.30	5.62*	5.77
fl417	8.36	3.99	1.96*
pcb442	10.89	3.51*	3.93
u574	7.76	5.74*	7.14
p654	3.85*	6.69	7.82
rat783	8.28	5.76*	6.27
pr1002	7.83	5.17*	6.66
u1060	7.48	6.34*	6.58
pcb1173	9.86	7.35	5.30*
d1291	13.61	6.06	4.88*
rl1323	11.87	4.89*	5.11
fl1400	6.94	5.39*	7.55
u1432	9.06	6.47	5.75*
fl1577	12.05	9.16	7.25*
d1655	12.24	6.26*	8.50
vm1748	8.18	5.71*	6.86
rl1889	11.42	6.75	6.71*
u2152	15.51	5.56*	6.95
pr2392	9.69	6.36*	7.41
pcb3038	8.76	5.99	5.60*
fl3795	10.21	6.30	5.54*
fnl4461	7.59	5.02*	6.57
rl5934	14.69	6.86*	8.69
Average	9.62	5.80	6.21

Table 7.14 Results of combined 2-opt/node insertion

Results on Table 7.14 display a substantial increase in quality compared to the pure 2-opt heuristic (cf. Table 7.8). Quality improves from 15% to 10% for random starts and from 9% to 6% for the other two variants.
For random instances, BENTLEY (1992) reports performance of 6.7% above an approximation of the Held-Karp bound for a combination of 2-opt and node insertion.
CPU times (given in Figure 7.15) are higher by a factor of about two. Again, complete search is usually not of practical relevance and we incorporate the candidate subgraph consisting of the 10 nearest neighbor graph and the Delaunay graph. Only those 2-opt or node insertion moves that introduce at least one edge of this graph into the tour are considered.

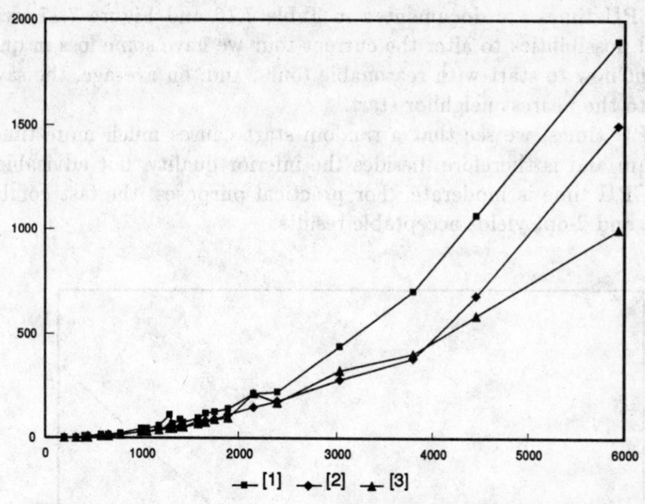

Figure 7.15 CPU times for combined 2-opt/node insertion

Problem	Random	Nearest N.	Savings
d198	3.42*	9.06	4.73
lin318	5.93	5.74*	6.01
fl417	10.44	8.50	2.43*
pcb442	8.80	3.95	3.93*
u574	7.19	7.12	7.08*
p654	37.81	27.47	6.98*
rat783	8.24	7.44	6.33*
pr1002	11.53	7.65	6.31*
u1060	11.75	8.14	6.31*
pcb1173	9.71	7.17	5.20*
d1291	17.82	8.69	4.95*
rl1323	16.30	8.85	5.23*
fl1400	14.80	13.55	8.69*
u1432	9.55	6.81	5.77*
fl1577	40.27	18.55*	19.10
d1655	12.54	7.91	8.58*
vm1748	9.28	5.88*	6.98
rl1889	15.33	7.22	7.10*
u2152	17.82	9.00	7.14*
pr2392	11.97	7.90	7.28*
pcb3038	8.71	5.92	5.96*
fl3795	54.15	15.63*	21.38
fnl4461	8.30	4.94*	6.56
rl5934	18.05	10.25	8.75*
Average	15.40	9.31	7.45

Table 7.16 Results of fast combined 2-opt/node insertion

Results and CPU times are documented in Table 7.16 and Figure 7.17, respectively. Due to limited possibilities to alter the current tour we have some loss in quality. It is more important now to start with reasonable tours, and, on average, the savings start is preferrable to the nearest neighbor start.
Looking at CPU times, we see that a random start causes much more time to reach a local optimum and is therefore, besides the inferior quality, not advisable. For the other starts, CPU time is moderate. For practical purposes, the fast combination of node insertion and 2-opt yields acceptable results.

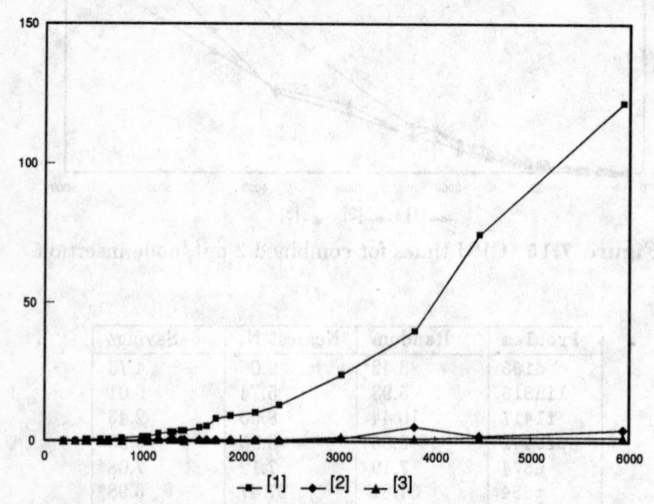

Figure 7.17 CPU times for fast combined 2-opt/node insertion

At the end of this section we display a tour for problem rd100 that admits no more 2-opt or node insertion moves. This tour has length 8647, and is therefore 9.3% above the optimal value 7910.
There are still further possibilities to speed up the 2-opt heuristic. In our enumeration scheme, the nodes of the problem are stored in a list (initialized according to the sequence of the nodes in the tour). In every iteration step the first node is taken from the list, scanned as described above, and reinserted at the end of the list. If i is the current node to be scanned we examine if we can perform an improving 2-opt move which introduces a candidate edge having i as one endnode. If an improving move is found then all four nodes involved in that move are stored at the beginning of the node list (and therefore reconsidered with priority). The crucial point for speeding up computations further is to reduce the number of distance function evaluations which accounts for a large portion of the running time. A thorough discussion of this issue can be found in BENTLEY (1992).
One idea is to avoid reevaluation of the same moves. Namely, when considering a candidate edge $\{i, j\}$ for taking part in a 2-opt move, we check if i and j have the same neighbors in the tour as when $\{i, j\}$ was considered previously. In that case no 2-opt move involving $\{i, j\}$ can be successful.

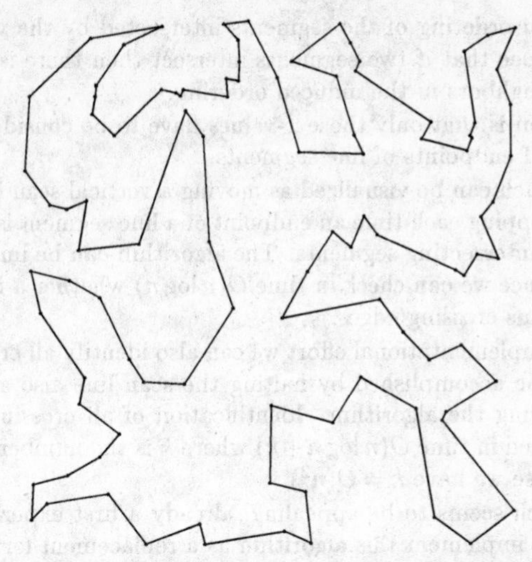

Figure 7.18 A tour optimal with respect to 2-opt and node insertion

In addition, one can perform a limited range search in the following sense. We only consider $\{i,j\}$ if this edge is not longer than the edges from i to both of its neighbors in the tour. The heuristic is run until the search for an improving 2-opt node fails for every node.
A further general observation is the following. Usually, decrease in the objective function value is considerable in the first steps of the heuristic while it is "tailing off" later. In particular, it takes a final complete round through all allowed move possibilities to verify that no further improving move is possible. Therefore, if one stops the heuristics early (e.g., if only a very slow decrease is observed over some period) not too much quality is lost. In our experiments, appropriate limits for the number of examined nodes were $3n$ or $4n$.

7.3 Crossing Elimination

In the Euclidean case we have seen that every pair of crossing tour edges gives rise to an improvement possibility. While, in general, it takes time $O(n^2)$ to check all pairs of edges we can take advantage of the Euclidean case to identify tour crossings faster.
We apply methods from computational geometry for finding intersections of line segments. The algorithm we are going to outline uses the so-called **scan line** principle or **plane sweep** approach. We only describe the algorithm for testing if there exists a crossing at all (for details see OTTMANN & WIDMAYER (1990)).
The algorithm uses a basic observation. Let A and B be two line segments (tour edges) in the plane and let x be a horizontal coordinate. We say that A is above B with respect to x if the vertical line with horizontal coordinate x intersects both A and B and if the intersection point with A lies above the intersection point with B. For any fixed x this

definition induces an ordering of the segments intersected by the vertical line through $(x, 0)$. It is easy to see that if two segments intersect then there is an x such that the two segments are neighbors in the induced ordering.
A further observation is that only those x-values have to be considered which are horizontal coordinates of endpoints of line segments.
The scan line approach can be visualized as moving a vertical scan line across the plane from left to right stopping each time an endpoint of a line segment is hit. At such points we look for pairs of intersecting segments. The algorithm can be implemented to run in time $O(n \log n)$. Hence we can check in time $O(n \log n)$ whether a tour for a Euclidean TSP instance contains crossing edges.
With a little more implementational effort we can also identify all crossing pairs quickly. Basically, this can be accomplished by halting the scan line also at every intersection point computed during the algorithm. Identification of all crossing pairs of segments can then be performed in time $O(n \log n + k)$ where k is the number of such pairs. Note that in the worst case we have $k = O(n^2)$.
Though the approach seems to be appealing, already a first experiment shows that it is not worthwhile to implement this algorithm as a replacement for the 2-opt heuristic. Figure 7.19 shows a tour for problem kroB100 which contains no crossings, but still is 82.9% above optimum.

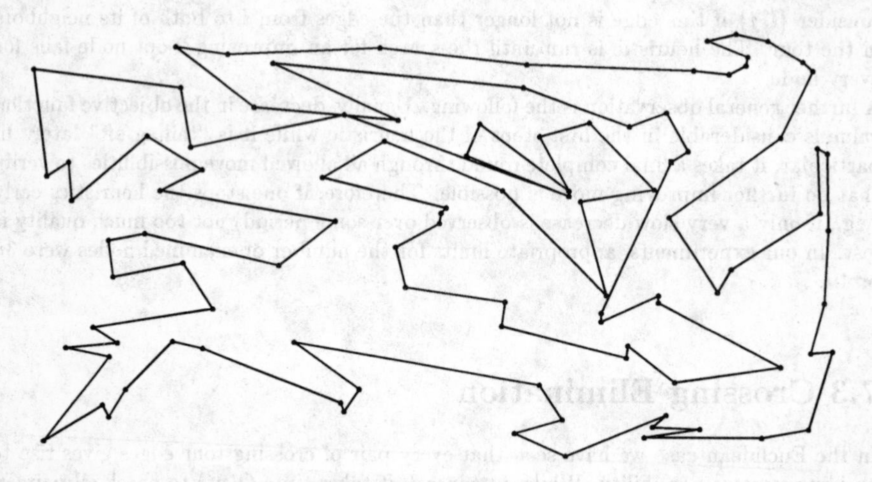

Figure 7.19 A noncrossing tour for kroB100

Nevertheless, we can use this special crossing elimination for Euclidean TSPs as a preprocessing step before using the 2-opt or another local improvement heuristic. We have not elaborated on this.

7.4 The 3-Opt Heuristic and Variants

A possible improvement of the 2-opt heuristic is obvious. To have more flexibility for modifying the current tour we could break the tour into three parts instead of only two and combine the resulting paths in the best possible way. Such a modification is called **3-opt move**. The number of combinations to remove three edges of the tour is $\binom{n}{3}$ and there are eight ways to connect three paths to form a tour (if each of them contains at least one edge).
Note that node insertion, edge insertion, and 2-opt exchange are special 3-opt moves. Node (edge) insertion is obtained if one path of the 3-opt move consists of just one node (edge). A 2-opt move is a 3-opt move where one eliminated edge is used again for reconnecting the paths.
Out of the eight 3-opt moves only four are real 3-opt moves introducing new tour modifications not captured by node/edge insertion or 2-opt moves.
To examine all 3-opt moves whether they can contribute to decrease the tour length takes time $O(n^3)$. Tour update after a 3-opt move is also more complicated than in the 2-opt case. The direction of the tour may change on all but the longest of the three involved paths.
Therefore, we decided to not consider a full 3-opt at all, but to limit in advance the number of 3-opt moves that are considered. To give an impression of the time needed for a full 3-opt: it takes about 3 hours to perform the 3-opt heuristic for problem **p654** starting with a nearest neighbor tour.
The implemented procedure is the following.

procedure 3-opt

(1) Let T be the current tour.

(2) For every node $i \in V$ compute a set of nodes $N(i)$ (possible choices are discussed below).

(3) Perform the following until failure is obtained.

(3.1) For every node $i = 1, 2, \ldots, n$:
Examine all possibilities to perform a 3-opt move which eliminates three edges having each at least one endnode in $N(i)$. If it is possible to decrease the tour length this way, then choose the best such 3-opt move and update T.

(3.2) If no improving move could be found, then declare failure.

end of 3-opt

If we limit the cardinality of $N(i)$ by some fixed constant independent of n, then checking in Step (3.2) if an improving 3-opt move exists at all takes time $O(n)$ (but with a rather large constant hidden by the O-notation).
We implemented the 3-opt routine by using a dynamic enumeration order and maintaining the direction of the tour on the longest path. The interval data structure was not used in the experiments. A further speed-up can be obtained for very large problems if it is employed.

We considered the following three different candidate sets:

1) the 10 nearest neighbor subgraph,
2) the 10 nearest neighbor subgraph augmented by the Delaunay graph,
3) a subgraph generated by the candidate heuristic discussed in section 5.3 with parameters $w = 15$ and $k = 10$.

For a given candidate subgraph G_C we used the neighborhood set $N(i)$ consisting of all neighbors of i in G_C. To limit CPU time (which is cubic in the cardinality of $N(i)$ for every node i in Step (3.1)) the number of nodes in each set $N(i)$ was bounded by 50. We applied the restricted 3-opt heuristic to the four starting tours:

1) random tour,
2) nearest neighbor tour,
3) savings tour,
4) Christofides tour.

The Christofides heuristic was executed in its simplified version as it is described in section 6.3. Table 7.20 shows the results with the first candidate set.

Problem	Random	Nearest N.	Savings	Christofides
d198	2.86	5.27	1.69	1.24*
lin318	2.93	2.80	2.16*	4.62
fl417	5.90	3.68	0.62*	5.24
pcb442	5.67	1.66*	2.54	3.03
u574	5.83	3.98	4.21	3.07*
p654	7.69	3.80	1.06*	5.78
rat783	4.60	3.47*	4.26	3.88
pr1002	4.45	3.61	4.02	3.24*
u1060	7.06	6.17	4.04	3.20*
pcb1173	5.82	5.70	3.93*	4.29
d1291	14.40	4.26	2.74*	6.30
rl1323	7.61	4.51	3.43	3.35*
fl1400	10.06	6.34	3.86*	7.20
u1432	7.47	4.83	3.51	3.36*
fl1577	15.85	8.40	8.71	4.89*
d1655	9.43	4.23*	5.09	5.17
vm1748	5.98	6.05	3.84	3.69*
rl1889	8.22	6.91	5.54	3.78*
u2152	9.82	5.78	4.85*	4.62
pr2392	7.15	4.37	4.94	3.62*
pcb3038	6.82	4.12	4.63	4.09*
fl3795	17.48	10.27	7.35	4.53*
fnl4461	4.77	3.45	5.13	3.40*
rl5934	14.39	6.37	6.74	4.11*
Average	8.01	5.00	4.12	4.15

Table 7.20 Results of 3-opt (Variant 1)

The savings and the Christofides starting tours lead to the best average results. This follows again our rule of thumb that all relatively simple heuristics profit from a starting tour that gives some guidelines on how a good tour should look like. On the other hand,

this starting tour should not be too good in the sense that improvements could only be achieved if the global structure of the tour is changed. Therefore, not always will the best starting tour lead to the best final result.
For their respective versions of 3-opt, an excess of 3.6% over an approximation of the Held-Karp bound is reported in JOHNSON (1990) and an excess of 4.5% was achieved in BENTLEY (1992).

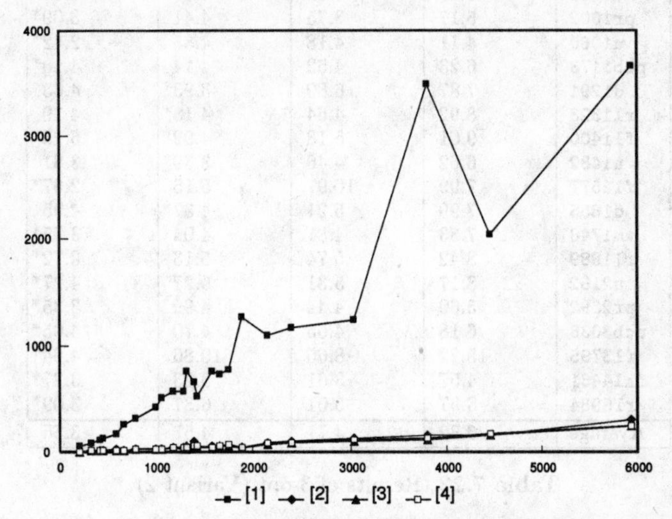

Figure 7.21 CPU times for 3-opt (Variant 1)

Figure 7.21 displays the CPU time for the 3-opt experiments of Table 7.20. Diagrams [1] to [4] corresponding to the starting tour variants 1) through 4). Except for the random starting tour, CPU times are moderate and similar for the other three starting tours. Practical application of this heuristic even for large problems is therefore possible. Observe again, that running time on real problems is not always well predictable. We have a outlier in case of the random start for problem fl3795 which took much more CPU time than it would have been expected from a simple extrapolation.
Tables 7.22 and Figure 7.23 display the results, CPU times resp., for the second candidate set.
Since this candidate set is larger and connected, one would expect a better performance than before. With the exception of the savings tour start, this is indeed the case. One generally observes, that it is hard to improve savings tours with simple exchanges. These tours are close to local optima, and there is a high chance that one gets stuck in a local optimum of inferior quality. Christofides' heuristic seems to be the construction procedure of choice in this case, giving solutions of average quality 3.55%. If we always take the better of the two solutions given in columns 4 and 5 of Table 7.22 we get an average quality of 3.23%. Due to the moderate CPU time requirements one could run both of these heuristics.
Finally, we employed our heuristic for finding suitable candidate edges. Table 7.24 and Figure 7.25 document the results and CPU times.

Problem	Random	Nearest N.	Savings	Christofides
d198	1.67	0.81*	3.41	1.10
lin318	4.23	2.74	2.18*	4.37
fl417	4.02	5.30	1.36	1.02*
pcb442	5.17	3.40	2.73	2.60*
u574	4.00	3.27	5.13	2.49*
p654	10.72	1.03*	4.84	4.11
rat783	5.46	4.41	4.58	4.01*
pr1002	5.17	3.75	4.41	3.09*
u1060	4.11	4.18	4.35	2.42*
pcb1173	6.23	4.52	4.14	4.10*
d1291	7.87	6.89	3.93*	4.63
rl1323	8.92	4.64	4.18*	4.19
fl1400	9.01	5.18	4.92*	5.23
u1432	6.62	4.46	3.39*	3.81
fl1577	7.99	10.97	5.15	2.87*
d1655	7.99	5.21	4.37*	4.95
vm1748	7.33	4.81	4.04	3.75*
rl1889	8.42	5.74	5.13	3.72*
u2152	8.17	5.31	5.27	4.27*
pr2392	5.60	4.44	4.99	3.35*
pcb3038	6.18	4.65	4.70	4.05*
fl3795	15.22	8.65	10.86	4.64*
fnl4461	4.57	3.61	5.23	3.27*
rl5934	8.97	5.01	6.37	3.09*
Average	6.82	4.71	4.57	3.55

Table 7.22 Results of 3-opt (Variant 2)

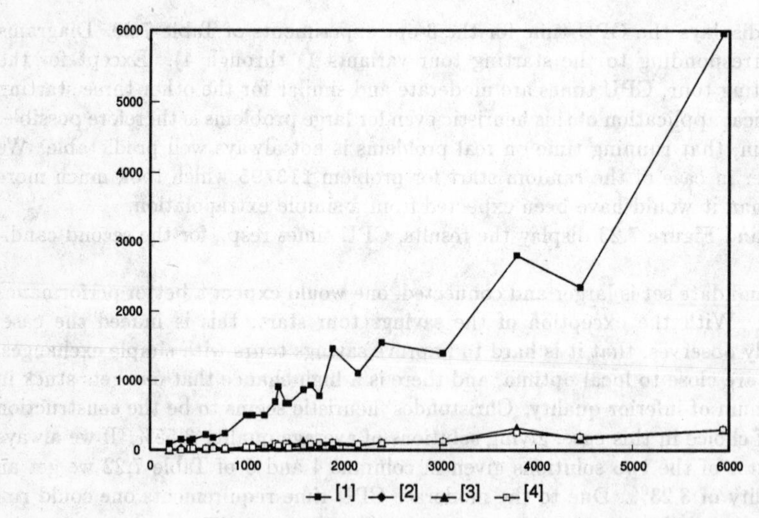

Figure 7.23 CPU times for 3-opt (Variant 2)

Problem	Random	Nearest N.	Savings	Christofides
d198	2.43	1.76	2.77	1.04*
lin318	2.32*	4.73	2.36	4.20
fl417	2.81	17.56	1.44*	2.70
pcb442	5.83	3.22	2.55*	2.69
u574	5.41	3.76	4.71	2.37*
p654	4.16	2.32*	5.57	4.59
rat783	5.44	3.12*	4.49	3.54
pr1002	6.92	6.05	6.05	3.96*
u1060	5.74	6.30	4.67	2.46*
pcb1173	6.96	7.00	5.43	4.07*
d1291	12.19	8.80	5.93*	6.32
rl1323	7.52	6.74	4.90	3.16*
fl1400	4.45*	4.78	5.04	5.18
u1432	7.68	6.16	5.04	4.13*
fl1577	12.58	8.28	13.28	4.76*
d1655	9.68	9.03	7.97	4.50*
vm1748	8.00	5.89	4.16	4.25*
rl1889	11.14	6.48	5.76	3.41*
u2152	9.03	6.00	5.94	5.51*
pr2392	8.64	5.86	6.89	3.77*
pcb3038	9.22	6.31	6.39	3.59*
fl3795	21.96	13.51	8.02	3.96*
fnl4461	9.14	5.46	6.12	3.80*
rl5934	15.71	8.49	7.53	3.52*
Average	8.12	6.57	5.54	3.81

Table 7.24 Results of 3-opt (Variant 3)

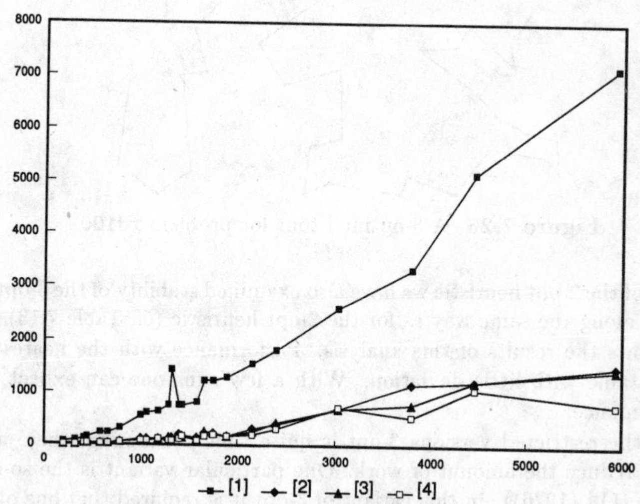

Figure 7.25 CPU times for 3-opt (Variant 3)

Qualities decreased with respect to the previous experiment. Only with the Christofides start, quality is comparable with the second candidate set. Note, however, that the Christofides heuristic requires the availability of a spanning tree. Whereas the Delaunay graph allows for an efficient computation of a minimum spanning tree, this is not possible with the third candidate set. Either, an efficient means for computing minimum spanning trees has to be available, or we cannot use the Christofides start with our candidate heuristic. An ordinary spanning tree computation would be too time consuming.

A further increase of the candidate set by specifying parameter values $w = 20$ and $k = 15$ lead to a further improvement giving solutions of average qualities 7.47%, 4.93%, 4.54%, and 3.5% resp., for the four different starting tours. However, it does not seem worthwhile to enlarge candidate sets too much because CPU time grows considerably.

Finally, in Figure 7.26, we show a 3-optimal tour for problem **rd100** which has length 8169 thus missing the optimal value by only 3.3%.

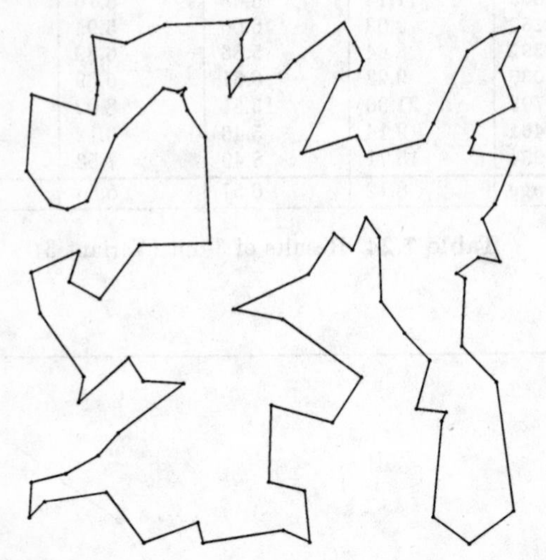

Figure 7.26 A 3-optimal tour for problem **rd100**

As in the case of the 2-opt heuristic we have also examined stability of the 3-opt heuristic. We proceeded along the same way as for the 2-opt heuristic (cf. Table 7.13).

Table 7.27 shows the results of this analysis. Performance with the nearest neighbor start is very stable with little deviation. With a few runs one can expect to achieve average performance.

Since also in the restricted versions 3-opt is quite time consuming, there are further approaches to reduce the amount of work. One particular variant is the so-called **Or-opt** procedure (OR (1976)). In this variant of 3-opt it is required that one of the paths involved in the move has a fixed prespecified length l. For example, edge insertion is then Or-opt with $l = 1$, and node insertion corresponds to $l = 0$. Results obtained with this procedure are not of high quality, so we do not go into details here.

Variant	Minimum	Maximum	Average	Span	Deviation
rd100					
1	0.00	12.11	3.48	12.11	2.17
2	0.85	6.31	3.15	5.46	1.36
d198					
1	0.32	11.02	2.98	10.70	2.08
2	2.12	7.67	7.02	5.55	1.04
lin318					
1	0.67	7.71	4.43	7.04	1.31
2	1.74	5.76	3.47	4.02	1.14
fl417					
1	0.51	32.75	9.11	32.24	5.38
2	5.57	6.42	6.21	0.84	0.08
pcb442					
1	2.09	8.06	4.93	5.97	1.04
2	1.37	3.26	2.11	1.89	0.37
u574					
1	1.95	10.80	5.10	8.85	0.97
2	2.09	4.48	3.24	2.39	0.52

Table 7.27 Stability analysis for the 3-opt heuristic

One might suspect that with increasing k the k-opt procedure should yield provably better approximate solutions. However, it is proved in ROSENKRANTZ, STEARNS & LEWIS (1977) that for every $n \geq 8$ and every $k \leq n/4$ there exists an instance of the TSP and a k-optimal tour such that

$$\frac{c_{k-\text{opt}}}{c_{\text{opt}}} = 2 - \frac{2}{n}$$

where $c_{k-\text{opt}}$ and c_{opt} are the respective values of the k-optimal and the optimal tour. Nevertheless, this is only a worst case result. We have seen in this section that it pays off to consider larger values of k and design efficient implementations of restricted k-opt procedures. The problem of designing efficient update procedures for k-opt heuristics is addressed in MARGOT (1992).

7.5 Lin-Kernighan Type Heuristics

The final heuristic to be discussed in this chapter was originally developed by LIN & KERNIGHAN (1973). The motivation for this heuristic is twofold and is based on experience we gained from the experiments described in the preceding sections of this chapter.

(i) The more flexible and powerful the possible tour modifications are, the better results are usually obtained (2-opt supersedes node insertion, 3-opt supersedes 2-opt, etc.).

(ii) Simple moves quickly get stuck in local optima of only moderate quality that cannot be left anymore.

The consequence of simply applying k-opt for larger k cannot be realized due to increasing running time. A complete check of the existence of an improving k-move takes time $O(n^k)$. One can, of course, also design restricted searches for several values of k, but we have not examined this feature. Rather, we overcome this drawback by using the ideas of Lin and Kernighan.
Their approach is based on the observation that sometimes a modification slightly increasing the tour length can open up new possibilities for achieving considerable improvements afterwards. The basic principle is to build complicated tour modifications that are composed of simple moves where not all of these moves necessarily have to decrease the tour length. To obtain reasonable running times the effort to find the parts of the composed move has to be limited.
Many variants of this principle are possible (see also MAK & MORTON (1993)). We did not use the original version of Lin and Kernighan, but implemented an own version where the basic components are 2-opt and node insertion moves. When building a move, we have in each substep some node from which a new edge is added to the tour according to some criterion. We illustrate our procedure by an example. Suppose we start with the canonical tour $1, 2, \ldots, 16$ for a problem on 16 nodes and we decide to construct a tour modification starting from node 16.
In the first step it is decided to eliminate edge $\{1, 16\}$ and to introduce the edge from node 16 to node 9 into the tour. Adding this edge creates a short cycle, and therefore edge $\{9, 10\}$ has to be deleted. To obtain a tour again, node 10 has to be connected to node 1.

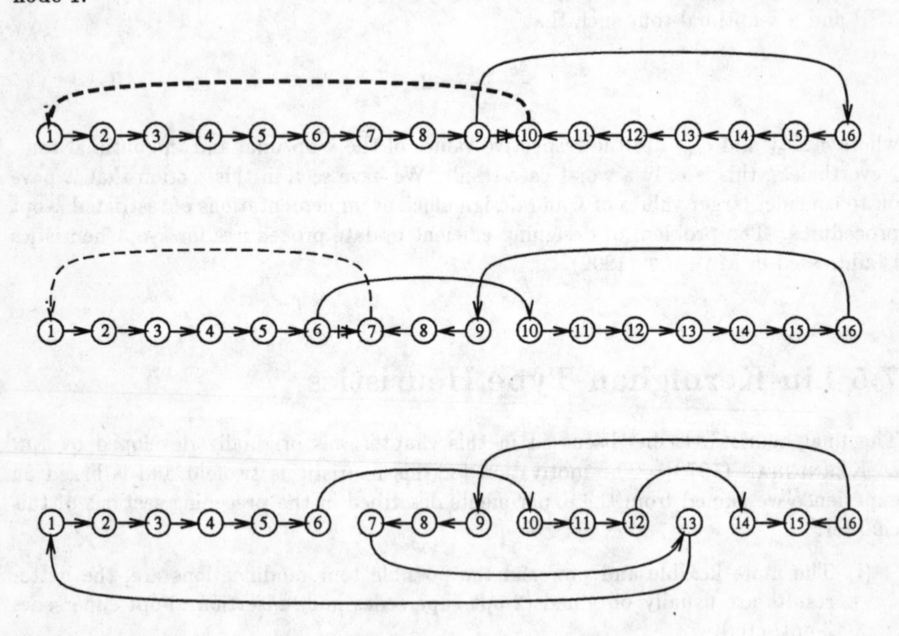

Figure 7.28 Lin-Kernighan moves

If we stop at this point, we have simply performed a 2-opt move. The fundamental new idea is now not to connect node 10 to node 1, but to search for another move starting from node 10. Suppose we now decide to add edge $\{10, 6\}$ to the tour. Again, one edge, namely $\{6, 7\}$ has to be eliminated to break the short cycle. The sequence of moves could be stopped here, if node 7 is joined to node 1.
As a final extension we instead add edge $\{7, 13\}$ and now perform a node insertion move by removing edges $\{12, 13\}$ and $\{13, 14\}$, adding edge $\{12, 14\}$, and connecting nodes 13 and 1 to form a tour.
Note that the direction on some parts of the tour changes while performing these moves and that this new direction has to be registered in order to be able to perform the next moves correctly. Building the final move, we obtained three different tours on the way. The shortest of these tours can now be chosen as the new current tour. It is not guaranteed (or required) that this is the final tour.
Realization of this principle is possible in various ways. We have chosen the following options.

(i) Submoves are 2-opt moves and optionally also node insertion moves.

(ii) To speed up search for submoves a candidate subgraph was used. Edges to be added from the current node to the tour were only taken from this set.

(iii) Edges to be added from the current node were selected according to the following local gain criterion. Let i be the current node. The local gain g_{ij} achieved by adding edge ij to the tour is computed as follows.
   - Let jk be the edge to be deleted if a 2-opt move is to be performed. Then $g_{ij} = c_{jk} - c_{ij}$.
   - Let jk and jl be the edges to be deleted if a node insertion move is to be performed. Then $g_{ij} = c_{jk} + c_{jl} - c_{lk} - c_{ij}$.

   The edge with the maximal local gain is chosen to enter the tour and the corresponding move is performed.

(iv) The number of submoves in a move is limited in advance.

(v) Examination of more than one candidate edge to enter the tour is possible. The maximal number of candidates examined from the current node and the maximal number of submoves up to which alternative edges are taken into account are specified in advance. This corresponds to inserting a limited enumeration component to the heuristic.

(vi) The interval data structure and a balanced binary search tree were used to store the tentative moves.

The basic outline of the heuristic is then given as follows.

procedure lin-kernighan

(1) Let T be the current tour.

(2) Perform the following until failure is obtained.

(2.1) For every node $i = 1, 2, \ldots, n$:

Try to find an improving move (consisting of 2-opt and node insertion submoves) starting from i according to the guidelines and parameters discussed above. If such an improving move can be found, then update T.

(2.2) If no improving move could be found, then declare failure.

end of lin-kernighan

A dynamic enumeration scheme using a priority queue is used to determine the starting node for the next move in step (2.1).

Possible choices of the parameters for this heuristic are so manifold that we cannot document all experiments here. We restrict ourselves to discuss some basic insight we obtained when conducting many series of evaluations.

In Tables 7.29 through 7.31 we display detailed results with several variants of the Lin-Kernighan exchange heuristic. All variants use 2-opt as well as node insertion submoves. They differ in the following respects.

1) Up to three submoves are considered for the first submove of a move. The number of submoves is limited to 15.
2) For the first three submoves of a move up to two alternatives are examined. This way up to eight moves are considered starting at node i selected in step (2.1). The number of submoves is limited to 15.
3) Same as 2), but the number of submoves building a move is now limited to 100.

Problem	Random	Nearest N.	Savings	Christofides
d198	0.75*	5.55	1.48	1.03
lin318	1.68	2.48	1.64	1.44*
fl417	3.10	0.61*	1.29	2.74
pcb442	1.49	1.95	2.33	1.33*
u574	1.96	2.48	2.11	0.93*
p654	1.71	3.07	0.05*	1.12
rat783	2.07	2.03*	2.92	3.21
pr1002	3.06	2.69	2.77	2.30*
u1060	2.88	2.41	2.73	1.78*
pcb1173	2.87	2.64	2.65	2.41*
d1291	4.77	4.51	2.97*	3.26
rl1323	2.25	2.68	1.79*	2.90
fl1400	2.30*	3.16	3.89	5.14
u1432	2.34	2.29	2.04*	2.20
fl1577	6.38	10.90	6.27	2.29*
d1655	2.63*	3.54	3.85	3.27
vm1748	2.11	2.00*	2.80	2.60
rl1889	2.46	3.45	3.90	2.35*
u2152	3.88	3.00	4.33	2.49*
pr2392	3.17	3.05	3.15	2.04*
pcb3038	2.50	1.82*	2.67	2.58
fl3795	3.46*	6.81	3.55	3.64
fnl4461	2.06	1.98*	2.47	2.26
rl5934	3.27	2.39	3.55	2.40*
Average	2.71	3.23	2.80	2.40

Table 7.29 Results of Lin-Kernighan (Variant 1)

We applied these three variants on the same four starting tours as in the case of the 3-opt heuristic, namely

1) random tour,
2) nearest neighbor tour,
3) savings tour,
4) Christofides tour.

The candidate subgraph consisted of the 10 nearest neighbor subgraph augmented by the Delaunay graph.

Problem	Random	Nearest N.	Savings	Christofides
d198	1.12	1.62	0.60*	1.10
lin318	2.00	1.22*	2.54	1.24
fl417	3.33	3.68	0.51*	3.00
pcb442	2.44	2.82	1.42*	1.90
u574	1.83	1.49*	2.20	2.32
p654	0.40	2.85	0.39*	2.50
rat783	2.76	2.23	2.80	1.90*
pr1002	2.68	3.20	2.66*	2.70
u1060	2.52	1.64	2.17	1.64*
pcb1173	2.90	1.77*	2.84	2.16
d1291	3.79	1.95*	3.14	4.93
rl1323	1.53	1.89	1.00*	2.97
fl1400	2.76*	3.49	3.20	2.92
u1432	1.74	1.53	2.00	1.47*
fl1577	3.63	5.67	6.58	1.48*
d1655	3.23	3.79	4.18	2.85*
vm1748	2.58	2.48	2.22*	2.25
rl1889	2.98	2.62	3.45	2.18*
u2152	2.67*	3.26	4.34	3.00
pr2392	2.35	2.02*	3.13	2.39
pcb3038	2.51	2.12	2.45	2.05*
fl3795	4.69	3.98	9.22	2.78*
fnl4461	1.72*	1.98	2.11	1.82
rl5934	3.40	3.49	2.92	2.03*
Average	2.56	2.62	2.84	2.32

Table 7.30 Results of Lin-Kernighan (Variant 2)

Tables 7.29, 7.30, and 7.31 represent only a small part of the experiments we have conducted with the Lin-Kernighan heuristic. In general, our observations can be summarized as follows.

- The more effort is spent, the better the results. (This is not a trivial statement, as the results in Chapters 6 and 7 show.)
- At least 15 submoves should be allowed for every move.
- It is better not to start out with a random tour, but to use locally good tours containing some major errors. But, this difference decreases with more elaborate versions of the Lin-Kernighan procedure.

Problem	Random	Nearest N.	Savings	Christofides
d198	2.81	0.75	0.60*	1.10
lin318	1.54	2.55	2.42	0.69*
fl417	0.75	3.75	0.51*	3.00
pcb442	2.12	1.39	1.30	1.11*
u574	2.74	1.59*	2.03	1.83
p654	0.59*	3.13	0.25	2.46
rat783	1.57*	1.71	2.80	1.79
pr1002	2.22	2.27	2.44	1.71*
u1060	1.76	1.69	2.09	1.54*
pcb1173	2.95	2.61	2.43	1.97*
d1291	3.81	2.13*	3.14	2.19
rl1323	2.13	2.43	1.00	1.97*
fl1400	3.58	3.04*	3.09	3.11
u1432	2.12	2.52	1.57*	1.76
fl1577	4.39	6.70	4.40	1.33*
d1655	2.64*	3.48	4.04	3.05
vm1748	1.80	1.92	2.27	1.73*
rl1889	2.34	2.93	2.96	1.62*
u2152	3.80	2.37	4.30	2.10*
pr2392	2.30	2.27	3.07	2.15*
pcb3038	1.99	2.31	1.90	1.85*
fl3795	4.84	3.04	3.67	2.81*
fnl4461	1.88	1.84	1.85	1.81*
rl5934	2.55	3.36	2.74	2.22*
Average	2.47	2.57	2.37	1.95

Table 7.31 Results of Lin-Kernighan (Variant 3)

- It is advisable to consider several alternate edges to be added from the first node.
- Inclusion of insertion moves usually leads to better results.

The results document, that, in contrast to simpler heuristics, the dependence on the starting tour is not very strong. Results and CPU times differ only slightly for the various starting tours. Again, it can be observed that not the best starting tours necessarily lead to the best final results. Our variant of the Christofides heuristic seems to be an appropriate choice. If one does not spend any effort at all to construct a starting tour (i.e., chooses a random starting tour) not much quality is lost, if at all.

We give an impression on the running times for the Lin-Kernighan heuristic in Figure 7.32. Diagrams [1] through [6] correspond to

1) Variant 1 with random start,
2) Variant 1 with Christofides start,
3) Variant 2 with random start,
4) Variant 2 with Christofides start,
5) Variant 3 with random start,
6) Variant 3 with Christofides start.

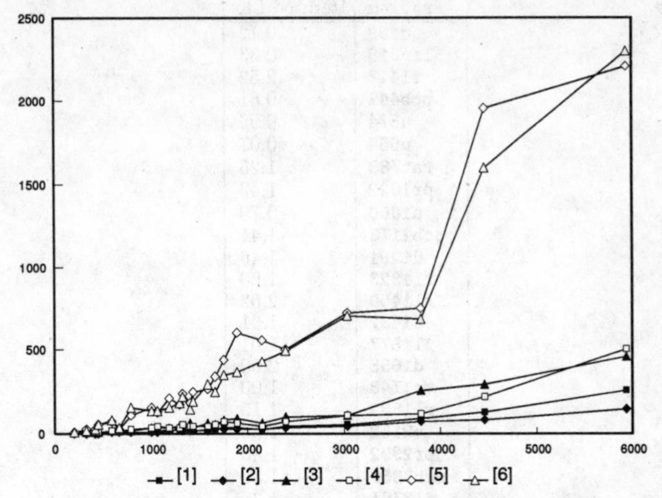

Figure 7.32 CPU times for Lin-Kernighan variants

With the Lin-Kernighan heuristic, the starting tour has only a minor effect on the CPU time. In particular, starting with a random tour now leads to about the same overall CPU time as starting with a more sophisticated tour. The reason for this can be seen when visualizing the Lin-Kernighan moves on a graphic screen. Even, when starting with a random tour, after execution of relatively few tour modifications, already a tour results that is of quality comparable to the quality of the best starting heuristics.

As every improvement heuristic, also the Lin-Kernighan heuristic has the problem of running into a local optimum that it cannot escape from. In contrast to the other improvement heuristics discussed before, the Lin-Kernighan has the feature that it can generate intermediate submoves that are not improving. This is in some sense a way to leave a local optimum. Each composite move, however, has to be improving.

A straightforward idea to escape from a local optimum that can be applied to every heuristic is the following. After the heuristic has found its approximate tour, we perturb the current tour by a random modification and then restart the heuristic. This seems of particular interest for the Lin-Kernighan heuristic since it finds very good local optima. For the other heuristics, this approach does not seem to be worthwhile. They terminate with approximate solutions that are relatively far away from the optimum, and several restarts will only slightly improve the quality.

We end this section reporting about a final experiment with this so-called **iterated Lin-Kernighan heuristic** which was first discussed in JOHNSON (1990). Using this idea, several optimal solutions of larger problems (e.g., pr2392) could be found. We chose Variant 3 and iterated it 20 times in the following way. Every time the heuristic terminated, a random 4-opt move was generated that did not increase the tour length by more than 10% and the heuristic was restarted. Recall that a 4-opt move consists of removing four edges from the current tour and patching together the resulting four paths in a different way.

Problem	Iterated LK
d198	0.12
lin318	0.33
fl417	2.52
pcb442	0.61
u574	0.96
p654	0.07
rat783	1.25
pr1002	1.43
u1060	0.79
pcb1173	1.41
d1291	1.46
rl1323	1.00
fl1400	2.62
u1432	1.31
fl1577	1.01
d1655	2.00
vm1748	1.00
rl1889	1.15
u2152	1.66
pr2392	1.46
pcb3038	1.36
fl3795	1.38
fnl4461	1.33
rl5934	2.10
Average	1.26

Table 7.33 Results of the iterated Lin-Kernighan heuristic

Results for just one starting tour, namely the Christofides tour, are depicted in Table 7.33. We see that further improvement in the quality of the final tour is achieved. Of course, running time is considerable and the iterated Lin-Kernighan heuristic is mainly suited for finding very good approximate solutions if enough CPU time can be spent.

7.6 Comparison of Improvement Heuristics

We conclude this section by comparing all improvement heuristics discussed in this chapter.

We apply the same method as in Chapter 6 to assess the relative quality of the improvement methods. For every problem instance, qualities are computed with respect to the best tour found by any of the methods. In addition we give the number of best solutions found by every heuristic. The comparison is given in Table 7.34. It should be easy to identify the listed heuristics. We have given the respective starting tour in parentheses. Furthermore, it is indicated if complete enumeration of all possible moves was performed or which candidate subgraph was used.

The figures in Table 7.34 should reflect reasonably well what can be expected from the various heuristics. In particular, if one wants to have solutions at most 1–2% above optimality one has to implement a Lin-Kernighan type improvement heuristic.

Heuristic	No. of best solutions	Relative quality
Iterated Lin-Kernighan	18	0.10
LK (10nn+Del, Christofides, Var. 3)	–	0.78
LK (10nn+Del, Christofides, Var. 2)	1	1.14
LK (10nn+Del, savings, Var. 3)	–	1.19
LK (10nn+Del, Christofides, Var. 1)	1	1.23
LK (10nn+Del, random, Var. 3)	–	1.29
LK (10nn+Del, random, Var. 2)	–	1.39
LK (10nn+Del, NN, Var. 3)	–	1.40
LK (10nn+Del, NN, Var. 2)	–	1.44
LK (10nn+Del, random, Var. 1)	1	1.54
LK (10nn+Del, savings, Var. 1)	1	1.62
LK (10nn+Del, savings, Var. 2)	2	1.65
LK (10nn+Del, NN, Var. 1)	–	2.05
3-opt (10nn+Del, Christofides)	–	2.36
3-opt (cand.heu, Christofides)	–	2.62
3-opt (10nn, savings)	–	2.92
3-opt (10nn, Christofides)	–	2.96
3-opt (10nn+Del, savings)	–	3.37
3-opt (10nn+Del, NN)	–	3.50
3-opt (10nn, NN)	–	3.80
3-opt (cand.heu, savings)	–	4.33
2-opt/NI (complete, NN)	–	4.59
2-opt/NI (complete, savings)	–	4.99
3-opt (cand.heu, NN)	–	5.34
3-opt (10nn+Del, random)	–	5.59
2-opt/NI (10nn+Del, savings)	–	6.21
3-opt (10nn, random)	–	6.76
3-opt (cand.heu, random)	–	6.87
Node insertion (complete, savings)	–	6.96
Node insertion (10nn + Del, savings)	–	7.05
2-opt (complete, NN)	–	7.18
2-opt (complete, savings)	–	7.50
2-opt/NI (10nn+Del, NN)	–	8.06
2-opt/NI (complete, random)	–	8.35
Edge insertion (complete, savings)	–	8.48
2-opt (10n+Del, savings)	–	8.79
Edge insertion (10nn + Del, savings)	–	8.86
2-opt (10n+Del, NN)	–	9.21
2-opt (complete, random)	–	13.34
2-opt/NI (10nn+Del, random)	–	14.08
Node insertion (complete, NN)	–	15.25
Node insertion (10nn + Del, NN)	–	15.38
2-opt (10n+Del, random)	–	15.66
Edge insertion (complete, NN)	–	16.08
Edge insertion (10nn + Del, NN)	–	18.86
Edge insertion (complete, random)	–	78.87
Node insertion (complete, random)	–	94.86
Edge insertion (10nn + Del, random)	–	113.54
Node insertion (10nn + Del, random)	–	125.25

Table 7.34 Comparison of improvement heuristics

We think that for practical applications the methods presented in this chapter are basic ingredients. They can be tuned to the running time that is available and come up with reasonable solutions.
Further improvement methods have been designed. E.g., GENDREAU, HERTZ & LAPORTE (1992) and GLOVER (1992) discuss additional types of exchange moves. The effect of the choice of the starting tour on the final result of exchange heuristics is considerd in PERTTUNEN (1991). Recently, stochastic improvement methods became rather popular. We will survey some of these approaches in Chapter 9.
Chapter 8 will present some further heuristics particularly suited for very large problems that will use the heuristics of this chapter for treating subproblems.

Chapter 8

Heuristics for Large Geometric Problems

In the previous chapters we have considered several heuristics for finding approximate solutions of traveling salesman problem instances. We have also shown how the use of candidate subgraphs can speed up computations enormously. There may be situations, however, where even these efficient implementations are too slow. If very restrictive real time constraints have to be observed, the methods derived so far may not be appropriate. In this chapter we address the question of finding traveling salesman tours for very large instances in short time. It is clear that we will have to accept some loss in quality.
We will consider some approaches for the fast determination of tours for 2-dimensional metric instances, i.e., instances defined by points in the 2-dimensional plane where the distances between points are given by some metric (recall that all our sample problem instances are Euclidean).
For the purposes of this chapter we have included some larger problems into our standard test set (and omitted the smallest problems). In particular, we added the real problems `rl5915`, `rl11849`, `brd14051`, and `d18512` as well as the random problem `rd15000`. As respective lower bounds we used the bounds 563416, 920847, 465044, and 644470 from TSPLIB for the first four instances, and the bound 175204 obtained with a Lagrangean approach based on 1-trees (see Chapter 10) for `rd15000`.

8.1 Space Filling Curves

We will begin our discussion of TSP heuristics with a fast heuristic suited only for problem instances in the Euclidean plane. The heuristic was given by BARTHOLDI & PLATZMAN (1982) and has some interesting theoretical properties.
Assume that all points are located in the unit square (which can always be achieved by suitable scaling) and that we want to find a short tour with respect to the Euclidean distance. The heuristic is based on a surjective mapping $\psi : [0,1] \to [0,1] \times [0,1]$, a so-called **space filling curve**. The name comes from the fact that when varying the arguments of ψ from 0 to 1 the function values fill the unit square completely. Surprisingly, such functions exist and, what is interesting here, they can be computed efficiently and also for a given $y \in [0,1] \times [0,1]$ a point $x \in [0,1]$ such that $\psi(x) = y$ can be found quickly.
We will describe the function used by Bartholdi and Platzman, which is recursively defined as follows.

$$\psi(x) = \begin{cases} \frac{1}{2}\psi(4x + \frac{1}{2} - \lfloor 4x + \frac{1}{2} \rfloor) & \text{if } 0 \leq x < \frac{1}{8} \text{ or } \frac{7}{8} \leq x \leq 1, \\ \frac{1}{2}[\psi(4x + \frac{1}{4} - \lfloor 4x + \frac{1}{4} \rfloor) + \binom{0}{1}] & \text{if } \frac{1}{8} \leq x < \frac{3}{8}, \\ \frac{1}{2}[\psi(4x - \lfloor 4x \rfloor) + \binom{1}{1}] & \text{if } \frac{3}{8} \leq x < \frac{5}{8}, \\ \frac{1}{2}[\psi(4x + \frac{3}{4} - \lfloor 4x + \frac{3}{4} \rfloor) + \binom{1}{0}] & \text{if } \frac{5}{8} \leq x < \frac{7}{8}. \end{cases}$$

Looking more closely at this expression, one realizes that it models the recursive subdivision of squares into four equally sized subsquares. The space filling curve is obtained by patching the four respective subcurves. Observe the rotation operation in the above formula which is necessary for making the subcurves fit smoothly.
The function ψ has interesting properties that are useful for the traveling salesman problem.

Theorem 8.1 *Let d_2 denote the Euclidean distance and define $f : [0,1] \to \mathbf{R}$ by $f(x) = 2\sqrt{x}$, if $0 \leq x \leq \frac{1}{2}$, and $f(x) = 2\sqrt{1-x}$, if $\frac{1}{2} < x \leq 1$.*

(i) Given $y \in [0,1] \times [0,1]$ in k-bit binary representation an $x \in [0,1]$ satisfying $\psi(x) = y$ can be computed in time $O(k)$.

(ii) For every $0 \leq x_1, x_2 \leq 1$ we have $d_2(\psi(x_1), \psi(x_2)) \leq f(|x_1 - x_2|)$.

(iii) If $0 \leq x_1 \leq x_2 \leq x_3 \leq 1$ then $f(x_3 - x_2) + f(x_2 - x_1) \geq f(x_3 - x_1)$.

(iii) If $0 \leq x_1 \leq x_2 \leq x_3 \leq x_4 \leq 1$ then $f(x_3-x_1)+f(x_4-x_2) \geq f(x_2-x_1)+f(x_4-x_3)$ and $f(x_3 - x_1) + f(x_4 - x_2) \geq f(x_3 - x_2) + f(x_4 - x_1)$.

□

For a set of numbers $x_i \in [0,1]$ the function f can be used to bound the length of a tour through the $\psi(x_i)$'s from above. Moreover, the best such bound is obtained if the points $\psi(x_i)$ are connected according to increasing values x_i. Based on this observation the space filling curves heuristic is given as follows.

procedure spacefill

(1) Scale the points to the unit square.

(2) For every scaled point $x_i = (x_i^1, x_i^2), i = 1, 2, \ldots, n$ compute z_i such that $\psi(z_i) = (x_i^1, x_i^2)$.

(3) Sort the numbers z_i in increasing order.

(4) Connect the points by a tour according to the sorted sequence of the z_i's (to complete the tour connect the two points with smallest and largest z-value).

end of spacefill

Due to Theorem 8.1 (i) and the time complexity $\Theta(n \log n)$ to sort the z_i's, this heuristic runs in time $\Theta(n \log n)$.
Figure 8.1 shows the result of this heuristic for a problem defined on the grid points of a 30×30 grid. The figure exhibits the recursive nature of the space filling curve.

Figure 8.1 A space filling curves tour

Suppose $z_{i_1}, z_{i_2}, \ldots, z_{i_n}$ is the sorted sequence computed in Step (3) of the heuristic. Because of Theorem 8.1 (ii) the length of the tour obtained can be bounded above by $\sum_{k=2}^{n} f(z_{i_k} - z_{i_{k-1}}) + f(z_{i_n} - z_{i_1})$. Since f is concave the maximum is attained for equally spaced arguments, i.e., $z_{i_k} - z_{i_{k-1}} = \frac{1}{n}$ (note that $f(z_{i_n} - z_{i_1}) = f(1 + z_{i_1} - z_{i_n})$). Hence a bound on the tour length is $n \cdot f(\frac{1}{n}) = 2\sqrt{n}$. In general, if the points are contained in a rectangle of area F, then the tour is not longer than $2\sqrt{nF}$. Bartholdi and Platzman have shown in addition that the quotient of the length of the heuristic curve and the shortest tour length is bounded by $O(\log n)$.

At this point we would like to comment briefly on average case analysis for the Euclidean TSP. Suppose that the n points are uniformly distributed in the unit square. BEARDWOOD, HALTON & HAMMERSLEY (1959) show that there exists a constant C such that

$$\lim_{n \to \infty} \frac{c_{\mathrm{opt}}}{\sqrt{n}} = C$$

where c_{opt} is the length of an optimal tour and give the estimate $C \approx 0.765$.

Such a behaviour can also be proved for the space filling curves heuristic with a different constant $\overline{C}$. BARTHOLDI & PLATZMAN (1982) give the estimate $\overline{C} \approx 0.956$. Therefore, for this class of random problems the space filling curves heuristic can be expected to yield tours that are approximately 25% longer than an optimal tour as n tends to infinity.

Table 8.2 shows the results of this heuristic and the CPU time for our set of sample problems. Since the space filling curves tour does not depend on the concrete configuration of the points (adding or deleting points does not change the relative order of the other points in the tour) it cannot be expected to perform too well in practice. In fact, our results only give an average quality of 46.8%.

Problem	Quality	CPU time
p654	44.29	0.05
rat783	42.99	0.07
pr1002	42.89	0.08
u1060	59.03	0.09
pcb1173	39.05	0.09
d1291	60.78	0.11
rl1323	62.28	0.11
fl1400	35.81	0.10
u1432	25.68	0.11
fl1577	69.48	0.13
d1655	44.98	0.14
vm1748	43.39	0.14
rl1889	57.58	0.15
u2152	50.11	0.17
pr2392	36.82	0.19
pcb3038	32.69	0.24
fl3795	68.40	0.30
fnl4461	33.71	0.36
rl5915	60.60	0.49
rl5934	58.80	0.48
rl11849	50.79	1.00
brd14051	36.53	1.16
rd15000	33.72	1.26
d18512	32.92	1.58
Average	46.80	

Table 8.2 Results of space filling curves heuristic

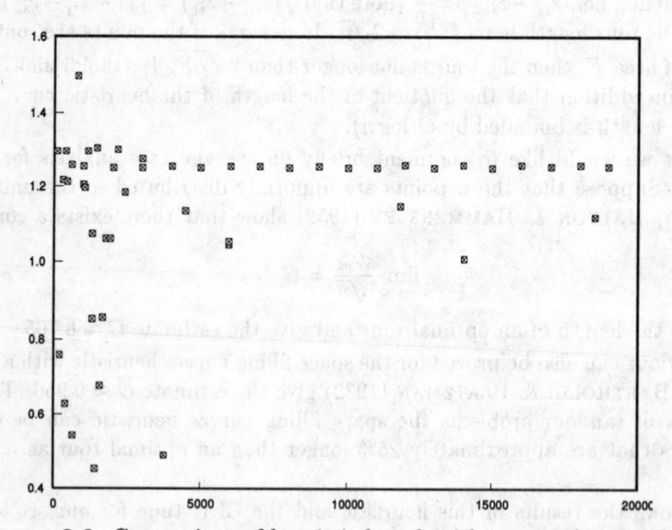

Figure 8.3 Comparison of heuristic length with expected optimal length

In Figure 8.3 we have depicted the quotients $c_{SF}/(0.765\sqrt{n})$, i.e., the deviation of the space filling curves tour from the expected optimal tour length, for a larger set of problem instances (note that $F = 1$ in our case). There is a considerable deviation of this quotient for real world problems, whereas for random problems instances the quotient is quite stable. In this range we have approximately $c_{SF}/(0.765\sqrt{n}) = 1.26$. This verifies that for points drawn at random from a uniform distribution in the unit square the heuristic yields tours about 25% above optimum. The same should be true for dense point sets equally distributed in the plane. Figure 8.3 also shows that the value $0.765\sqrt{n}$ (note that $F = 1$ in our case) is a very poor estimate for the optimal tour length of practical problem instances. For problem f13795 already the space filling curves heuristic gives a value of $0.376\sqrt{n}$.

8.2 Strip Heuristics

This is a well-known approach for finding tours for large geometric TSPs with little effort. The problem area is cut into vertical strips. Then a tour is constructed as follows. We start at the point in the first strip with lowest vertical coordinate, proceed to the node of strip 1 with next higher vertical coordinate, etc., until all points of the first strip are collected. The point with highest vertical coordinate in strip 1 is then connected to the node with highest vertical coordinate in strip 2, and strip 2 is scanned downwards. All remaining strips are then scanned connecting the points in the obvious way. Connecting the final point to the first point yields a Hamiltonian tour. Figure 8.4 shows a tour for problem pr1002 obtained by a strip heuristic.

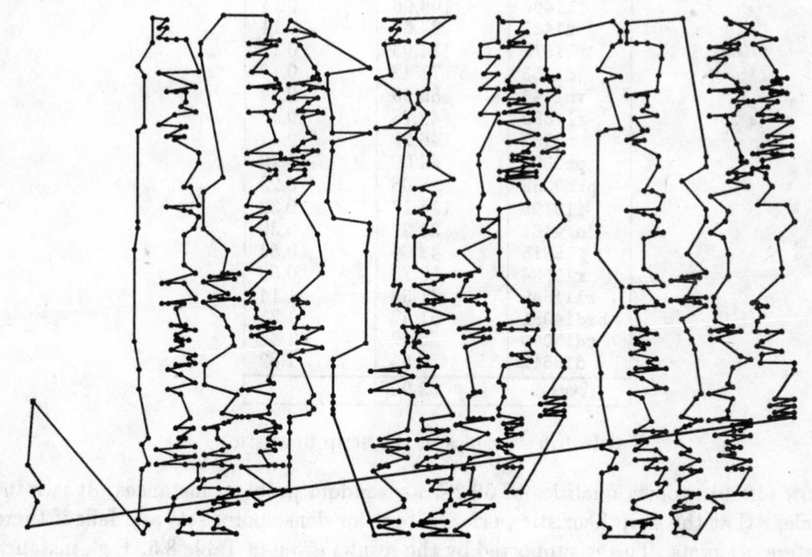

Figure 8.4 Tour for pr1002 found by a strip heuristic

As well as cutting the point area using vertical cuts, we could also use horizontal cuts. Furthermore, one can alternatively use cuts that result in strips of equal width or one could cut in such a way, that all strips contain the same number of points. Depending on the point configuration rather different results can arise. We used the following implementation.

procedure strip(s)

(1) Apply the strip heuristic in the following four cases and choose the best tour.
(1.1) Cut into s vertical strips of equal width.
(1.2) Cut into s vertical strips each containing the same number of points.
(1.3) Cut into s horizontal strips of equal height.
(1.4) Cut into s horizontal strips each containing the same number of points.

end of strip

The heuristic requires sorting the points with respect to horizontal and vertical coordinates. Since tour construction itself can then be done in linear time, the resulting running time is $\Theta(n \log n)$.
With a value of $s = \sqrt{n}/2$ we obtained the results of Table 8.5.

Problem	Quality	CPU time
p654	47.01	0.05
rat783	29.23	0.06
pr1002	52.74	0.08
u1060	53.81	0.08
pcb1173	27.37	0.09
d1291	76.51	0.10
rl1323	48.54	0.10
fl1400	106.06	0.11
u1432	34.65	0.11
fl1577	114.93	0.13
d1655	71.42	0.13
vm1748	56.39	0.14
rl1889	49.73	0.15
u2152	46.21	0.17
pr2392	42.00	0.19
pcb3038	27.68	0.25
fl3795	114.12	0.32
fnl4461	35.27	0.39
rl5915	42.66	0.53
rl5934	52.70	0.53
rl11849	41.33	1.14
brd14051	51.43	1.35
rd15000	33.42	1.46
d18512	39.65	1.82
Average	53.95	

Table 8.5 Results of the strip heuristic

JOHNSON (1990) reports qualities of 30.2% for random problem instances. It is intuitively clear that the strip heuristic performs best for dense point sets and fails if there are clusters of points. This is supported by the results given in Table 8.5. E.g., instance rd15000 is a random instance, whereas fl3795 contains clusters of points.

8.3 Partial Representations

The aim of this heuristic is to achieve a reduction of the number of nodes of the problem in such a way that the remaining nodes still give a satisfactory representation of the geometry of the original points. Then a traveling salesman tour on this set of representative nodes is computed in order to serve as a global approximation of the tour. In the final step the original nodes are inserted into this tour (where the number of insertion points that will be checked can be specified) and the representative nodes (if not original nodes) are removed. More precisely, we use the following bucketing procedure.

procedure partial_representation(m, k, l)

(1) Start with the rectangle enclosing the points whose vertices are determined by the minimal and maximal horizontal and vertical coordinates of the points.

(2) Recursively subdivide each rectangle into four equally sized parts by a horizontal and a vertical line until each rectangle
- contains no more than 1 point, or
- is the result of at least m recursive subdivisions and contains no more than k points.

(3) Represent each (nonempty) rectangle by the center of gravity of the points contained in it.

(4) Compute a tour through the representative nodes using some TSP heuristic.

(5) Insert the original points into this tour. To this end at most $l/2$ insertion points are checked before and after the corresponding representative nodes in the current tour. The best insertion point is then chosen.

(6) Remove all representative nodes that are not original nodes.

end of partial_representation

The parameters m, k, and l, and the heuristic needed in Step (4) can be chosen with respect to available CPU time. If the points of the problem are independently identically distributed in the unit square according to a uniform distribution then the running time of the above procedure is approximately $O(n \cdot (\max\{m, \log \frac{n}{k}\} + l))$ plus the time spent in Step (4) for computing the global tour.
We demonstrate our bucketing procedure for the 654-city problem **p654**. Figure 8.6 shows the bucketing obtained by setting $m = 3$ and $k = 25$. Figure 8.7 displays the 57 representative nodes which still contain much information about the geometric structure of the original problem. A global tour was computed in this case using the nearest neighbor heuristic to compute a starting tour and then performing 2-opt moves to improve this tour. The resulting tour is displayed in Figure 8.8. Subsequently, the feasible solution shown in Figure 8.9 for the original problem was obtained by inserting the original nodes into this tour considering at most 300 possible insertion points for each original node.

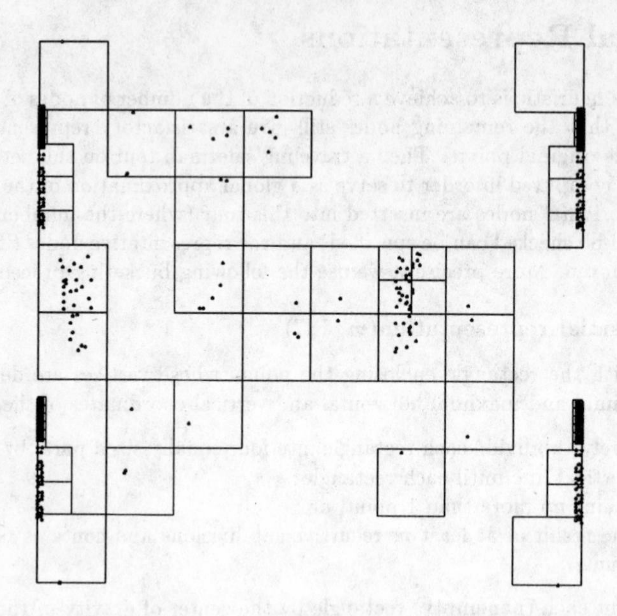

Figure 8.6 A bucketing for the problem p654

Figure 8.7 The corresponding set of representative nodes

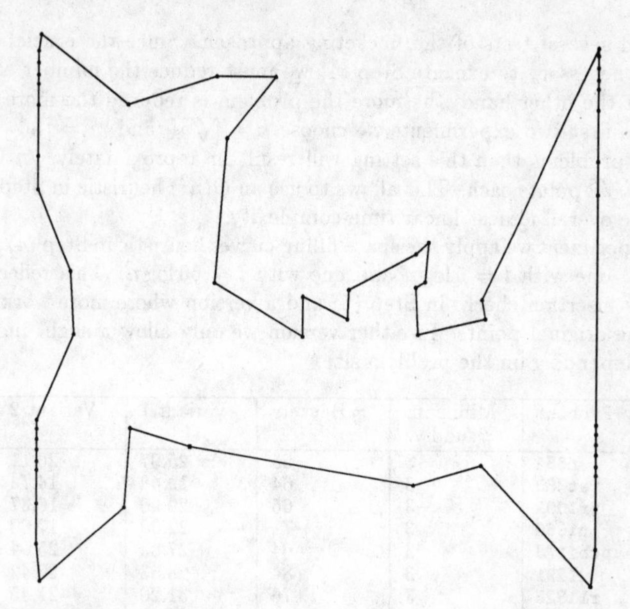

Figure 8.8 A global tour through the representative nodes

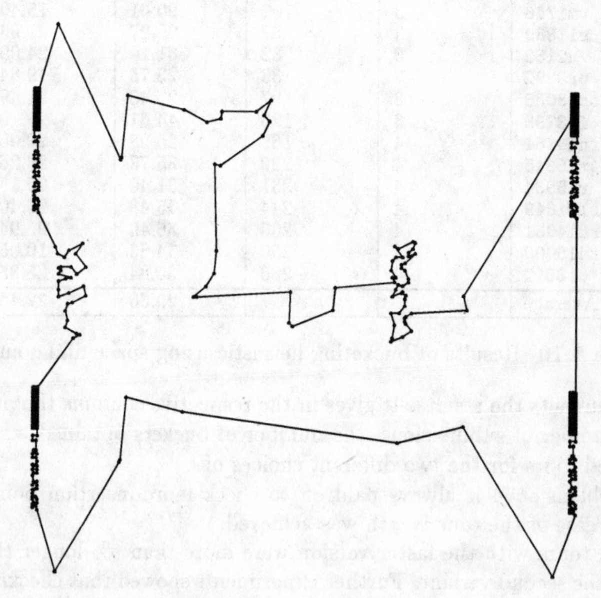

Figure 8.9 A feasible tour for the original problem

We performed several tests of the bucketing approach. Since the crucial point is the running time necessary to execute Step (4) we must reduce the number of nodes considerably. On the other hand, the more the problem is reduced the more information is lost. In the first two experiments we choose $k = \lceil\sqrt{n}\rceil$ and $m = \lceil\log_2 \sqrt[4]{n}\rceil$. If we have random problems then this setting will result in approximately $\sqrt{n}$ buckets containing about $\sqrt{n}$ points each. This allows to use an $O(n^2)$ heuristic in Step (4) without destroying the overall almost linear time complexity.
In the first experiment we apply the space filling curves heuristic in Step (4) and perform two test runs, one with $l = 5\log n$ and one with $l = 50\log n$. This reflects a version with very few insertion checks in Step (5) and a version where more work is spent for reinserting the original points. In either version we only allow a slight increase in the parameter l depending on the problem size.

Problem	Minimum #subdiv.	#Buckets	Variant 1	Variant 2
p654	3	55	25.97	26.11
rat783	3	64	19.48	14.74
pr1002	3	65	20.96	16.37
u1060	3	71	22.09	17.00
pcb1173	3	64	27.83	25.04
d1291	3	84	28.53	22.42
rl1323	3	76	31.20	21.45
fl1400	3	97	19.11	14.00
u1432	3	70	17.74	15.90
fl1577	3	82	41.39	38.97
d1655	3	104	27.87	25.41
vm1748	3	102	20.61	15.49
rl1889	3	87	29.27	21.93
u2152	3	82	31.10	24.09
pr2392	3	133	23.75	19.84
pcb3038	3	97	27.40	21.49
fl3795	3	126	44.61	29.18
fnl4461	4	169	25.98	22.05
rl5915	4	229	35.73	26.26
rl5934	4	231	31.40	25.13
rl11849	4	244	45.48	25.40
brd14051	4	265	36.49	25.94
rd15000	4	256	34.83	19.05
d18512	4	286	35.84	25.85
Average			29.36	22.46

Table 8.10 Results of bucketing heuristic using space filling curves

Table 8.10 documents the results. It gives in the respective columns the problem name, the minimal number of subdivisions, the number of buckets obtained, and the quality of the computed tours for the two different choices of l.
Except for problem p654 it always paid off to check more insertion points, usually a significant decrease of the tour length was achieved.
In the average, tours with the faster version were more than 6% longer than the tours obtained with the second variant. Further experiments showed that checking even fewer insertion points (say, for instance the constant number 20) leads to much inferior results.
In the next experiment we have put more effort in determining the global tour. To solve the global TSP we computed the subgraph of 6 nearest neighbors, computed a nearest

neighbor tour making use of this subgraph and performed the limited Lin-Kernighan heuristic where each move consisted of at most 15 node insertion or 2-opt moves. Table 8.11 gives the results. Parameter settings and columns are the same as for Table 8.10. The results show that not always better tours are obtained by computing better global tours. The difference in quality for the two experiments is not substantial.

Problem	Minimum #subdiv.	#Buckets	Variant 3	Variant 4
p654	3	55	7.32	7.64
rat783	3	64	20.11	19.96
pr1002	3	65	20.65	17.57
u1060	3	71	25.54	18.82
pcb1173	3	64	27.92	23.26
d1291	3	84	27.23	21.17
rl1323	3	76	31.74	22.93
fl1400	3	97	9.26	6.27
u1432	3	70	17.79	15.22
fl1577	3	82	21.61	17.97
d1655	3	104	23.91	20.08
vm1748	3	102	19.98	16.71
rl1889	3	87	30.16	23.93
u2152	3	82	28.74	25.98
pr2392	3	133	26.57	23.32
pcb3038	3	97	25.53	20.64
fl3795	3	126	31.28	20.87
fnl4461	4	169	24.59	21.79
rl5915	4	229	33.06	27.99
rl5934	4	231	29.99	25.14
rl11849	4	244	46.85	27.61
brd14051	4	265	36.13	25.39
rd15000	4	256	34.52	20.05
d18512	4	286	36.29	25.88
Average			26.53	20.67

Table 8.11 Results of bucketing heuristic using better global tours

It seems that the representation used was too coarse. As a final experiment we have therefore increased the number of buckets by setting $k = \lceil\sqrt{n}\rceil/2$. This way we obtain a finer representation of the original problem. We only got a slight increase (if at all) in the running times. Table 8.12 displays the results.

Results show that both having a fine representation and checking enough insertion points are important. Solutions get better if the representation is refined and if the number of insertion checks is increased. Of course, this is not surprising, because we approach the original problem this way. In any case running times grow. If we choose a relatively coarse representation and check many insertion points, then we will obtain nearest insertion like tours.

A graphical display of the running times for our experiments is given in Figure 8.13. Running times increase linearly with the problem size and can therefore be estimated quite precisely which can be important for some practical applications. To meet imposed timing restrictions the node reduction heuristic can be tuned by choosing the parameters m, k, and l and the heuristic in Step (4) appropriately.

Problem	Minimum #subdiv.	#Buckets	Variant 5	Variant 6
p654	3	124	6.85	6.84
rat783	3	76	21.85	18.88
pr1002	3	138	19.20	14.60
u1060	3	160	18.92	15.56
pcb1173	3	175	21.84	18.51
d1291	3	130	23.09	19.74
rl1323	3	177	24.69	20.17
fl1400	3	136	6.82	5.63
u1432	3	184	13.71	13.24
fl1577	3	139	18.59	15.13
d1655	3	190	22.23	20.38
vm1748	3	198	17.26	15.06
rl1889	3	197	19.21	14.65
u2152	3	235	25.04	22.70
pr2392	3	196	21.39	19.62
pcb3038	3	246	21.07	19.38
fl3795	3	247	23.97	18.44
fnl4461	4	238	23.32	20.42
rl5915	4	312	30.95	27.17
rl5934	4	315	27.26	25.48
rl11849	4	565	29.69	25.36
brd14051	4	574	27.18	23.44
rd15000	4	517	28.30	19.22
d18512	4	686	26.97	23.80
Average			21.64	18.48

Table 8.12 Results of bucketing heuristic with a finer representation

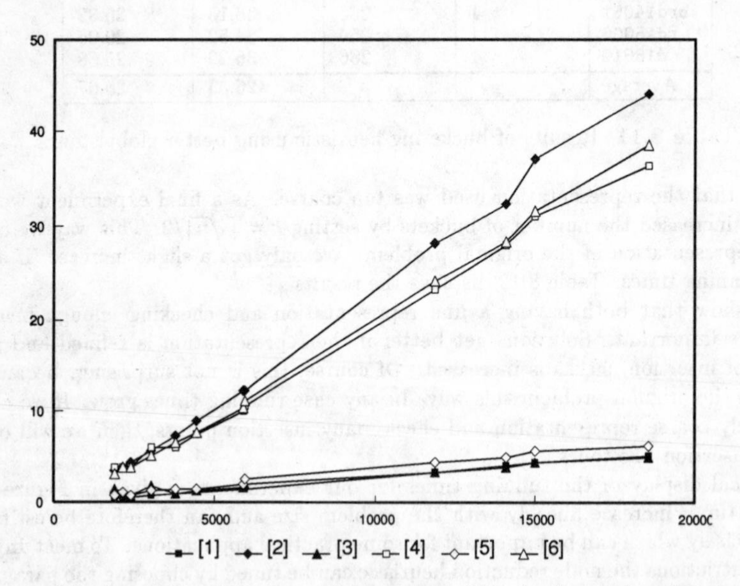

Figure 8.13 CPU times for variants of the bucketing heuristic

One might also think of exploiting parallelism. For example, the further subdivision of different rectangles can be performed in parallel (though the effect on the overall running time will only be small). Insertion of the original nodes can also be parallelized. Synchronization has to be observed here.

8.4 Decomposition Approaches

A further idea to reduce the complexity of a large scale problem instance is to partition it into subproblems, where the partitioning is done (if possible) in a way that structural properties of the problem instance are preserved.
We will examine three types of decompositions which are based on nearest neighbor computations, on the Delaunay graph and on a simple geometric partitioning scheme, respectively. In all cases we proceed as follows.

procedure decomposition

(1) Compute a partition of the points defining the problem instance into pairwise disjoint sets.

(2) Represent each subset by its convex hull, or more precisely by those points of the subset that are situated on the boundary of its convex hull.

(3) Compute a global tour connecting the subsets. To this end each subset is represented by a node. The distance between two such nodes is given by the shortest distance between original nodes that are located on the boundary of the convex hulls representing the respective subproblems. The global tour now gives an entering and a leaving point for each of the subproblems. If these two points are identical we apply a simple exchange to obtain different endpoints.

(4) Apply suitable TSP heuristics to find short Hamiltonian paths in every subproblem connecting the entering to the leaving point visiting all points of the subproblem.

(5) Merge the global tour and the paths to form a tour for the original problem.

end of decomposition

The running time of this procedure depends on the partition that was obtained and on the heuristics that are applied in Steps (3) and (4). We will see below that the partition itself can be obtained very fast in all cases. The heuristic is similar to clustering algorithm of LITKE (1984) where clusters of points are represented by a single point. Having computed an optimal Hamiltonian cycle through the representatives, clusters are expanded one after another. For Litke's clustering method, JOHNSON (1990) reports an average excess over the Held-Karp bound of 23.2% for randomly generated problem instances.

8.4.1 Nearest Neighbor Partition

We compute the 3 nearest neighbor subgraph. Since only few neighbors are taken into account, this subgraph will usually be highly disconnected. We take as a partition of the node set the partition induced by the connected components of this subgraph. If the partition is not appropriate (sometimes the 3 nearest neighbor partition has too few components to yield a reasonable partition.) we take the 2 nearest neighbor subgraph. The nearest neighbor partition can be computed in linear time from the Delaunay graph.

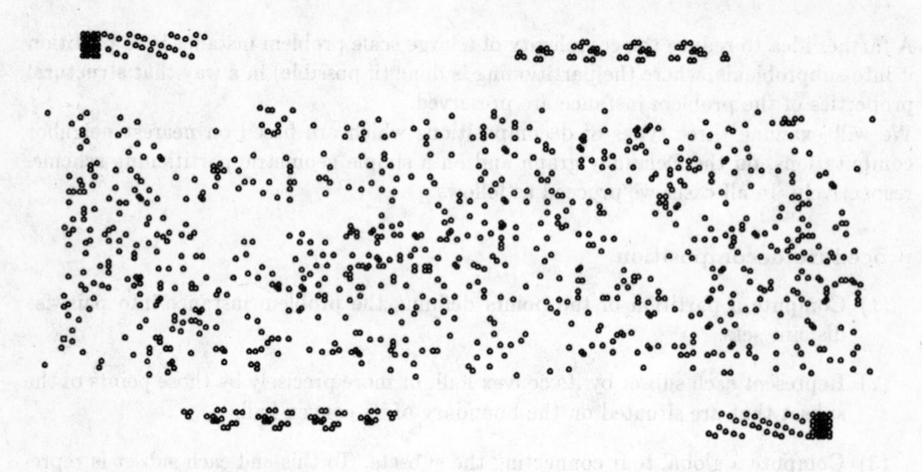

Figure 8.14 The traveling salesman problem u1060

Figure 8.15 A nearest neighbor partition for u1060

We consider as an example problem u1060 which is displayed in Figure 8.14 (for reasons of display, two points were eliminated). Figure 8.15 shows the partition given by the computation of the 2 nearest neighbors (subproblems are shown together with their respective convex hulls). Figure 8.16 displays the global tour (only the connections between the subproblems are shown) and the tour obtained for the original problem is given in Figure 8.17.

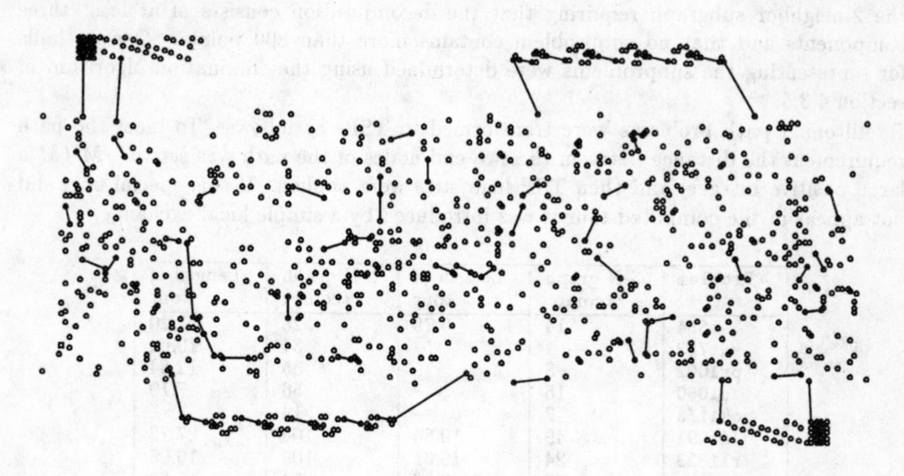

Figure 8.16 A global tour

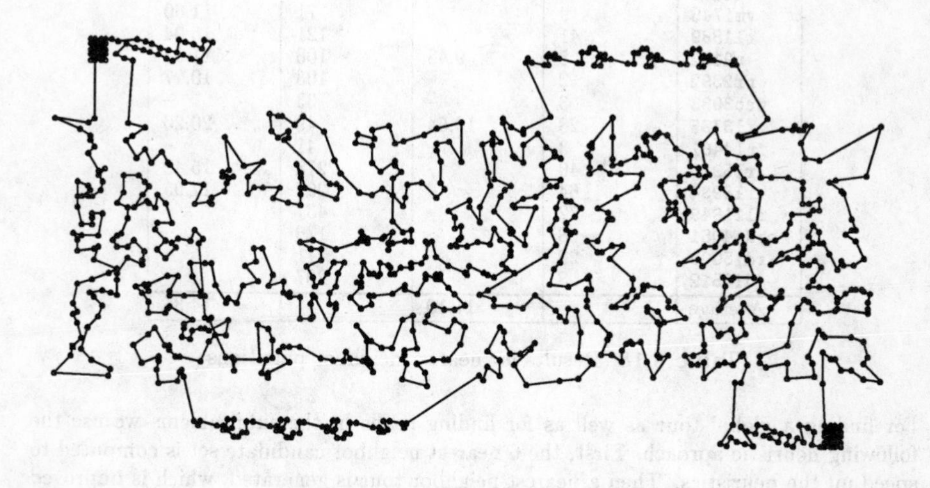

Figure 8.17 A feasible tour for u1060

It should be mentioned that the partitions do not always look as nice as in the above example. There are situations where convex hulls intersect or where one convex hull

is contained in another. But this does not necessarily influence the quality of the tour obtained. Also the sizes of the subproblems may be quite different. If the partition does not turn out to seem reasonable one should either try to change the partition or one should apply other heuristics. We will not elaborate on this here.
Experiments with the partitioning approach based on nearest neighbors were performed in the following way. We applied the heuristic to the 3-neighbor subgraph and to the 2-neighbor subgraph requiring that the decomposition consists of at least three components and that no subproblem contains more than 800 points. Convex hulls for representing the subproblems were determined using the elimination algorithm of section 4.3.5.
Hamiltonian path problems were transformed to TSPs as follows. To meet the path requirement the distance between the two end nodes of the path was set to $-M$ (M a large positive integer) and then TSP heuristics were applied. If this special edge did not appear in the computed tour it was introduced by a simple local exchange.

Problem	3-N graph #comp.	Length of tour	2-N graph #comp.	Length of tour
p654	14	5.78	25	7.20
rat783	1	–	34	10.64
pr1002	8	–	35	12.64
u1060	15	–	56	7.10
pcb1173	2	–	14	–
d1291	49	10.84	104	27.72
rl1323	24	16.61	109	16.96
fl1400	22	8.06	54	10.61
u1432	1	–	12	–
fl1577	19	12.61	31	24.05
d1655	20	5.82	53	9.67
vm1748	9	–	71	11.60
rl1889	41	–	121	16.04
u2152	55	9.48	108	11.68
pr2392	2	–	103	10.07
pcb3038	3	–	33	–
fl3795	28	19.64	49	20.26
fnl4461	3	–	41	–
rl5915	48	–	278	15.42
rl5934	64	–	297	17.95
rl11849	72	–	437	–
brd14051	8	–	176	–
rd15000	20	–	711	–
d18512	8	–	207	–
Average		11.10		14.35

Table 8.18 Results for nearest neighbor partitions

For finding a global tour as well as for finding paths in the subproblems we use the following heuristic aproach. First, the 6 nearest neighbor candidate set is computed to speed up the heuristics. Then a nearest neighbor tour is generated, which is improved by applying the Lin-Kernighan heuristic. In the Lin-Kernighan heuristic, each move is composed of at most 15 2-opt or node insertion submoves.
The computational results are displayed in Table 8.18. Missing entries are either due to the fact that too few subproblems were generated or that at least one subproblem

contained more than 800 points. In addition to the computed tour lengths for the two variants we have displayed the number of components in the respective partitions. Since many entries are missing, we do not give a figure of the CPU times here. E.g., it took 3.8, 7.3, and 11.9 seconds to perform the heuristic based on the 2 nearest neighbor graph for problems `pcn3038`, `fnl4461`, and `rl5915`, respectively.
Nearest neighbor decompositions seem suitable for clustered problems. If the points are equally distributed in the plane then this partition seems not to be appropriate.

8.4.2 Delaunay Decomposition

This partition also attempts to assign points that are very likely to be close in good tours to the same subset. Subsets obtained from nearest neighbors seem to be reasonable since in near-optimal tours many edges will connect nearby points. Also the Delaunay graph should exhibit relevant neighborhood relations between points. We compute a subgraph of the Delaunay graph as follows (l is the minimal number of connected components of this subgraph and s is the maximal size of a connected component).

procedure delaunay_partition(l, s)

(1) Sort the edges of the Delaunay graph with respect to increasing lengths.

(2) Examine the edges in this order and add an edge to the subgraph if
  - the number of connected components does not fall below l, and
  - no component's size is increased above s.

(3) Take the partition induced by the subgraph as the partition of the problem.

end of delaunay_partition

Due to the sorting Step (1) it takes time $\Omega(n \log n)$ to obtain the Delaunay partition. Note that this procedure cannot fail and always ends up with a partition meeting the requirements.
Two experiments with the Delaunay partition were performed.

(1) Here we require to have at least 30 components with at most 50 nodes in each component.

(2) The lower limit on the number of components is set to $\lceil \sqrt{n} \rceil$ and the maximal number of nodes in a component is set to $\lceil \sqrt{n} \rceil$.

The resulting subproblems were solved in the same way as for the nearest neighbor partition. Computational results are displayed in Table 8.19. CPU times are given in Figure 8.20.
Again, as in the case of all previous approaches there are many possibilities for conducting further experiments.

Problem	Partition 1 #comp.	Length of tour	Partition 2 #comp.	Length of tour
p654	30	4.58	27	19.31
rat783	30	11.13	35	14.93
pr1002	30	12.68	32	17.59
u1060	32	15.78	34	14.79
pcb1173	30	13.60	37	14.14
d1291	30	12.48	36	10.33
rl1323	30	13.60	42	18.95
fl1400	70	21.92	88	28.29
u1432	45	13.92	59	13.80
fl1577	33	18.93	45	22.11
d1655	35	19.61	44	20.62
vm1748	47	16.40	52	15.69
rl1889	37	16.00	44	21.02
u2152	44	15.39	47	14.66
pr2392	52	14.55	52	14.55
pcb3038	70	14.50	70	12.75
fl3795	123	32.51	114	27.62
fnl4461	137	13.55	116	14.05
rl5915	136	22.37	90	21.88
rl5934	141	23.39	96	20.74
rl11849	277	18.22	130	16.50
brd14051	423	13.48	266	15.16
rd15000	430	13.88	264	13.70
d18512	578	13.68	355	13.53
Average		16.09		17.36

Table 8.19 Results for Delaunay partitioning variants

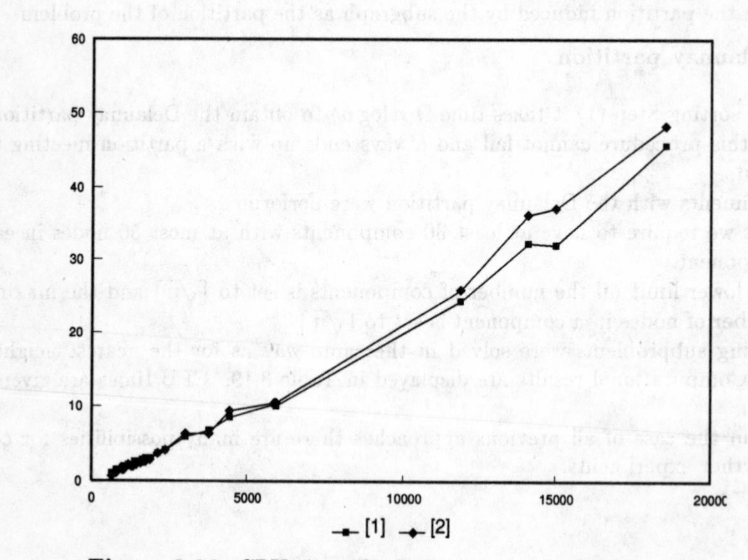

Figure 8.20 CPU times for Delaunay partitioning heuristics

8.4.3 Rectangle Decomposition

In KARP (1977) the problem area is divided by horizontal and vertical cuts such that each segment contains no more than a certain number k of points. Then, for each segment an optimal Hamiltonian cycle on the points of that segment is computed (using dynamic programming), and in a final step all subtours are glued together according to some scheme to form a Hamiltonian cycle through all points. For fixed k the optimal solutions of the respective subproblems can be determined in linear time (however, depending on k, a large constant associated with the dynamic programming part is hidden). Some interesting results on average behavior can be found in KARP AND STEELE (1985). A further decomposition approach is discussed in HU (1965).
We consider an even simpler geometric partitioning scheme. We enclose the problem area by the smallest rectangle parallel to the coordinate axes. Then this rectangle is divided by $k-1$ horizontal and $k-1$ vertical cuts into k^2 smaller rectangles of equal width and equal height. Every nonempty rectangle defines one subset of the partition. In our experiments, we have chosen $k = \sqrt[4]{n} - 1$ and $k = \sqrt[4]{n}$. For random problems, the choice $k = \sqrt[4]{n}$ will yield about $\sqrt{n}$ subproblems, each of size $\sqrt{n}$. Table 8.21 shows the results for the two parameter settings, Figure 8.22 displays the CPU times. The choice $k = \sqrt[4]{n} + 1$ gives similar results, therefore they are not listed here.

Problem	$k=\sqrt[4]{n}-1$	$k=\sqrt[4]{n}$
p654	11.10	7.74
rat783	9.19	11.78
pr1002	9.85	16.54
u1060	11.80	16.33
pcb1173	9.43	12.02
d1291	17.79	15.56
rl1323	14.70	17.93
fl1400	10.41	8.93
u1432	6.85	8.05
fl1577	18.48	17.93
d1655	6.71	14.13
vm1748	10.21	13.23
rl1889	15.71	14.73
u2152	13.04	14.71
pr2392	11.33	10.63
pcb3038	9.36	9.37
fl3795	13.89	12.09
fnl4461	8.01	8.67
rl5915	13.10	15.18
rl5934	13.48	13.00
rl11849	11.12	12.71
brd14051	7.61	7.80
rd15000	8.95	9.83
d18512	7.56	7.47
Average	11.24	12.35

Table 8.21 Results for rectangle decomposition

Results compare favorably with the nearest neighbor and the Delaunay partition. For the purpose of finding reasonable tours in very short time, this simple decomposition seems to be appropriate.

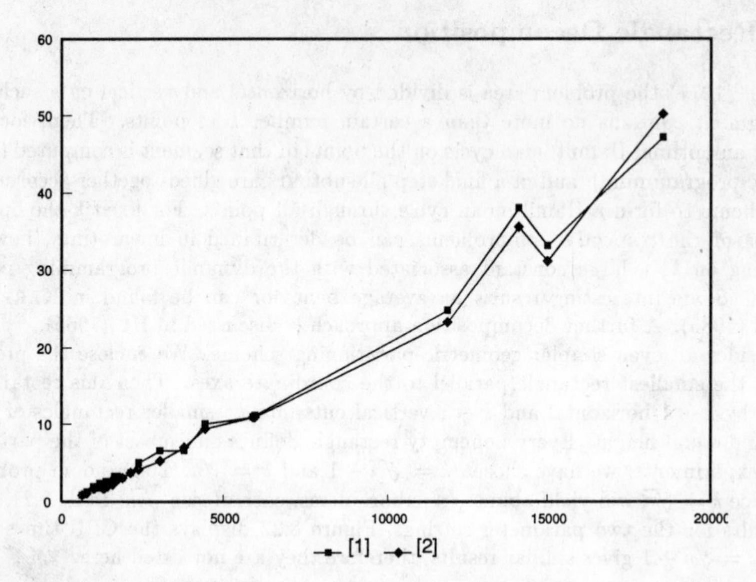

Figure 8.22 CPU times for rectangle decomposition heuristics

A particularly interesting property of partitioning heuristics is that solving the subproblems can be parallelized easily because they are completely independent. To come up with short tours, however, emphasis has to be given to find suitable partitions. More research has to be undertaken here.

It is also possible to derive some quality guarantee for the tours obtained this way. If we add up lower bounds for the length of the global tour and for the lengths of the Hamiltonian paths for the subproblems we obtain a lower bound on the length of tours that can be computed using this particular partition.

It should be emphasized that the heuristics of this chapter are meant to complement the well-known approaches for finding approximate TSP solutions. They are mainly useful for finding fast solutions in the range of 10–20% quality. In any case, decomposition seems to be the key concept for treating large problems in short time and for introducing parallelism to TSP solving. WOTTAWA (1991) studies the implementation of TSP heuristics on a parallel hardware architecture. Further research, in particular concerning decomposition approaches for finding optimal solutions, has to be conducted.

Chapter 9

Further Heuristic Approaches

Every TSP heuristic has, at least in principle, the chance to find an optimal tour. But, this is an almost impossible event. Usually, an improvement method only finds a locally optimal tour in the sense that, although the tour is not optimal, no further improving moves can be generated. The weaker the local moves that can be performed, the larger the difference between the length of an optimal tour and of the tour found by the heuristic. One way to obtain better performance is to start improvement heuristics many times with different starting tours, because this increases the chance of finding better local minima. Success is limited, though.
Another possibility is to perturb the current tour by some modification that increases the length of the tour and to restart the heuristic. The iterated Lin-Kernighan heuristic follows this principle by perturbing the locally optimal tour using a random 4-opt move. Most of the heuristics we will consider in this section try to use a systematic rule to escape from local minima or avoid local minima. A basic ingredient is the use of randomness which contrasts these approaches to purely deterministic heuristics. We have not implemented the various heuristics, but give references to the literature.

9.1 Simulated Annealing

The approach of **Simulated Annealing** originates from theoretical physics where Monte-Carlo methods are employed to simulate phenomena in statistical mechanics. Its predecessor is the so-called **Metropolis filter** (METROPOLIS, ROSENBLUTH, ROSENBLUTH, TELLER & TELLER (1953)). This simulation method can be motivated as follows. Consider a huge number of particles (e.g., gas molecules) of fixed volume at some temperature ϑ. Since the particles move, the system can be in various states. The probability that the system is in a state of certain energy E is given by the Boltzmann distribution $f(E) = \exp(-E/(\kappa_B\vartheta))/z(\vartheta)$, where $z(\vartheta)$ is a normalization factor and κ_B is the Boltzmann constant. This distribution characterizes the statistical equilibrium of the system at temperature ϑ. The physical system is simulated as follows.

procedure Metropolis_filter

(1) Generate an initial state given by the positions of the particles.

(2) Perform the following steps for a given number of iterations.

(2.1) Tentatively change the position of one particle by some random displacement and evaluate the resulting change ΔE of the energy of the system.

(2.2) Accept the new position if $\Delta E \leq 0$ and with probability $\exp(-\Delta E/(\kappa_B \vartheta))$ if $\Delta E > 0$.

end of Metropolis_filter

Interestingly, if the procedure is iterated long enough, then the distribution of the states it generates is indeed the Boltzmann distribution. The necessary number of iterations to reach this statistical equilibrium is called **relaxation time** and depends on the temperature ϑ. Relaxation time increases with decreasing temperature. For $\vartheta \to 0$ only states of minimal energy will receive a positive mass in the distribution.

Simulated annealing uses the Metropolis filter for several temperature steps. Consider as an analogy the physical process of cooling a liquid to its freezing point with the goal to obtain an ordered crystalline structure. Rapid cooling would not achieve this, one rather has to cool (anneal) the liquid very slowly in order to allow improper structures to readjust and to have a perfect order (ground state) at the crystallization temperature. At each temperature step the system relaxes to its statistical equilibrium.

KIRKPATRICK, GELATT & VECCHI (1983) connect such a physical process with an optimization method for a combinatorial minimization problem. Feasible solutions correspond to states of the system (an optimal solution corresponding to a ground state, i.e., a state of minimal energy). The objective function value resembles the energy in the physical system. System dynamics is imitated by random local modifications of the current feasible solution. Relaxation is modeled in that, depending on the level of the temperature, alterations that increase the energy (objective function) are more or less likely to occur. At low temperatures, it is very improbable that the energy of the system increases. Pure improvement heuristics as we have discussed so far can be interpreted in this context as rapid quenching procedures that do not allow the system to relax.

We give a general outline of a simulated annealing procedure for the TSP.

procedure simulated_annealing

(1) Compute an initial tour T and choose an initial temperature $\vartheta > 0$ and a repetition factor r.

(2) As long as the stopping criterion is not satisfied perform the following steps.

(2.1) Do the following r times.

(2.1.1) Perform a random modification of the current tour to obtain the tour T' and let $\Delta = c(T') - c(T)$ (difference of lengths).

(2.1.2) Compute a random number x, $0 \leq x \leq 1$.

(2.1.3) If $\Delta < 0$ or $x < \exp(-\Delta/\vartheta)$ then set $T = T'$.

(2.2) Update ϑ and r.

(3) Output the current tour T as solution.

end of simulated_annealing

Step (2.1) is essentially the Metropolis filter. The formulation has several degrees of freedom and various realizations are possible. Usually 2-opt or 3-opt moves are employed as basic modification in Step (2.1.1). The temperature ϑ is decremented in Step (2.2) by setting $\vartheta = \gamma\vartheta$ where $\gamma < 1$ is a real number close to 1, and the repetition factor r is usually initialized with the number of cities and updated by $r = \alpha r$ where α is some factor between 1 and 2. Realization of Step (2.2) determines the **annealing schedule** or **cooling scheme**. The scheme given above is named geometric cooling. The procedure is stopped, if the length of the current tour was not altered during several temperature steps. Figure 9.1 shows a typical run of a simulated annealing procedure. The x-axis counts the moves accepted, the y-axis gives the objective function value.

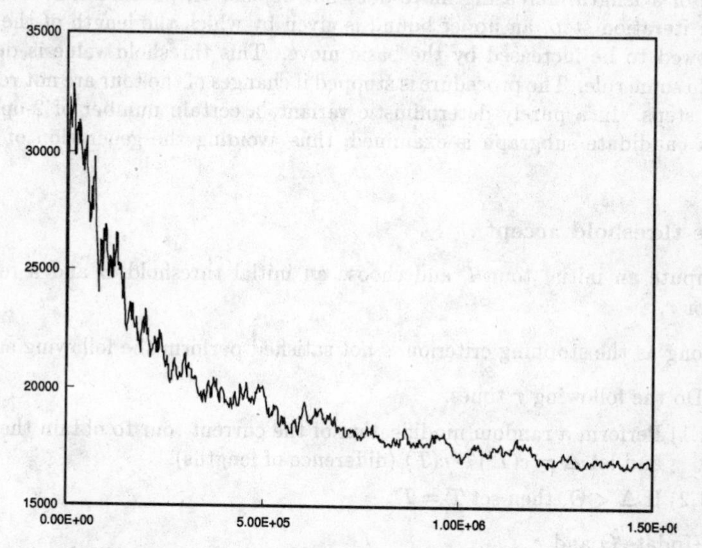

Figure 9.1 Example of a simulated annealing run

Expositions of general issues for the development of simulated annealing procedures are AARTS & KORST (1989a) and JOHNSON, ARAGON, MCGEOCH & SCHERON (1991), a bibliography is given in COLLINS, EGLESE & GOLDEN (1988). Computational experiments for the TSP are, for example, reported in KIRKPATRICK (1984), CERNY (1985), VAN LAARHOVEN (1988), AARTS & KORST (1989b), and JOHNSON (1990). It is generally observed that simulated annealing can find very good or even optimal solutions and beats Lin-Kernighan concerning quality. To be certain of this, however, one has to spend considerable CPU time because temperature has to be decreased very slowly and many repetitions at each temperature step are necessary.
We think that the most appealing property of simulated annealing is its fairly simple implementation. The principle can also be used to approach very complicated problems, since one only needs a basic subroutine that turns a feasible solution into another feasible solution. But, a requirement that should not be underestimated is the proper choice of the annealing scheme. This is highly problem dependent and only numerous experiments can find the most suitable parameters.

HAJEK (1985) proves the interesting theoretical results that, if the basic move satisfies some property and if a certain annealing schedule is used, this algorithm converges to an optimal solution with probability 1. Unfortunately, the theoretically required scheme is not suited for practical use, because it leads to very slow convergence.
A variant of simulated annealing enhanced by deterministic local improvement is discussed in MARTIN, OTTO & FELTEN (1992). When properly implemented, near optimal solutions can be found faster than with pure simulated annealing. A further heuristic motivated by phenomena from physics is **simulated tunneling** described in RUJÁN (1988).
DUECK & SCHEUER (1990) propose a simpler variant of simulated annealing, where acceptance of a length increasing move does not depend on probabilities. Rather, at each major iteration step, an upper bound is given by which the length of the current tour is allowed to be increased by the basic move. This threshold value is decreased according to some rule. The procedure is stopped if changes of the tour are not registered for several steps. In a purely deterministic variant, a certain number of 2-opt moves based on a candidate subgraph is examined, thus avoiding the generation of random moves.

procedure threshold_accept

(1) Compute an initial tour T and choose an initial threshold Θ and a repetition factor r.

(2) As long as the stopping criterion is not satisfied perform the following steps.

(2.1) Do the following r times.

(2.1.1) Perform a random modification of the current tour to obtain the tour T' and let $\Delta = c(T') - c(T)$ (difference of lengths).

(2.1.2) If $\Delta < \Theta$, then set $T = T'$.

(2.2) Update Θ and r.

(3) Output the current tour T as solution.

end of threshold_accept

Computational results display similar behaviour as simulated annealing. ALTHÖFER & KOSCHNICK (1989) give a theoretical convergence result.
An even simpler variant, called "**Great Deluge**" heuristic, is discussed in DUECK (1993). Here, at each major iteration step, there is an upper limit on the length of tours that are accepted. Every random move yielding a tour better than this length is accepted. The name of this approach comes from the interpretation that (for a maximization problem) the limit corresponds to the level of water which is rising during the heuristic and moves leading "into the water" are not accepted.

procedure great_deluge

(1) Compute an initial tour T and choose an initial upper limit U and a repetition factor r.

(2) As long as the stopping criterion is not satisfied perform the following steps.

(2.1) Do the following r times.

(2.1.1) Perform a random modification of the current tour to obtain the tour T'.

(2.1.2) If $c(T') \leq U$, then set $T = T'$.

(2.2) Update U and r.

(3) Output the current tour T as solution.

end of great_deluge

With fairly moderate computation time this method is reported to yield good results for practical traveling salesman problems arising when drilling printed-circuit boards.

9.2 Evolutionary Strategies and Genetic Algorithms

The following two related approaches were motivated by trying to imitate evolution in nature, which can find very good (or presumably optimal) solutions to highly complex problems.
The first approach is termed **evolutionary strategy** since it is based on analogues of "mutation" and "selection" to derive an optimization heuristic (RECHENBERG (1973)). Its basic principle is the following.

procedure evolution

(1) Compute an initial tour T.

(2) As long as the stopping criterion is not satisfied perform the following steps.

(2.1) Generate a modification of T to obtain the tour T'.

(2.2) If $c(T') - c(T) < 0$ then set $T = T'$.

(3) Output the current tour T as solution.

end of evolution

Note, that moves increasing the length of the tour are not accepted. The term "evolution" is used because the moves generated in Step (2.1) should be biased by knowledge aquired so far, i.e., somehow moves that have lead to a decrease of tour length should influence the generation of the next move. This principle, however, is hardly followed in practice, moves taken into account are usually k-opt moves generated at random. Formulated this way, the procedure cannot leave local minima and experiments show that it indeed gets stuck in rather poor local minima. Moreover, convergence is slow justifying the name "creeping random search" which is also used for this method.
To leave local minima one has to incorporate the possibility for perturbations that increase the tour length (ABLAY (1987)). Then this method can resemble a mixture of

pure evolutionary strategy, simulated anneling, threshold accept, and tabu search (see below).
More powerful in nature than just mutation-selection is genetic recombination. Interpretated in terms of the TSP this means that new tours should not be constructed from just one parent tour but rather be a suitable combination of two or more tours. Based on this principle, so-called **genetic algorithms** were developed.

procedure genetic_algorithm

(1) Compute an initial set $\mathcal{T}$ of tours.

(2) As long as the stopping criterion is not satisfied perform the following steps.

(2.1) Recombine two or more tours of $\mathcal{T}$ to obtain a new tour T which is added to $\mathcal{T}$.

(2.2) Reduce the set $\mathcal{T}$ according to some rule.

(3) Output the best tour found during the heuristic as solution.

end of genetic_algorithm

We see that Step (2.1) mimics reproduction in the population $\mathcal{T}$ and that Step (2.2) corresponds to a "survival of the fittest" rule.
Concrete realizations are manifold. Usually, subpaths of given tours are connected to form new tours and reduction is just keeping the set of k best tours of $\mathcal{T}$. One can also apply deterministic improvement methods to the newly generated tour T before performing Step (2.2). For an introduction, see GOLDBERG (1989). Applications to the TSP are reported in FRUHWIRTH (1987), MÜHLENBEIN, GORGES-SCHLEUTER & KRÄMER (1988), GORGES-SCHLEUTER (1990), and ULDER, PESCH, VAN LAARHOVEN, BANDELT & AARTS (1990). Findings of optimal solutions are reported for some problem instances with considerable amount of CPU time.

9.3 Tabu Search

Since the above heuristics allow length-increasing moves, local minima can be left during computation. No precaution, however, is taken to prevent the heuristic to revisit a local minimum several times. The approach of **tabu search** guarantees that it is forbidden (tabu) to return to the same feasible solution.

procedure tabu_search

(1) Compute an initial tour T and start with an empty tabu list $\mathcal{L}$.

(2) As long as the stopping criterion is not satisfied perform the following steps.

(2.1) Perform a move that is not forbidden by $\mathcal{L}$.

(2.2) Update the tabu list $\mathcal{L}$.

(3) Output the best tour found during the heuristic as solution.

end of tabu_search

Again, there are various possibilities to realize a heuristic based on the tabu search principle. Basic difficulties are the design of a reasonable tabu list, the efficient management of this list, and the selection of the most appropriate move in Step (2.1). GLOVER (1989) gives a detailed introduction to tabu search methods. KNOX & GLOVER (1989), MALEK, GURUSWAMY, OWENS & PANDYA (1989), and MALEK, HEAP, KAPUR & MOURAD (1989) report computational results for the TSP.

9.4 Neural Networks

The final heuristic is motivated by research on the functioning of the human brain. Basically, it models a set of neurons connected by a certain type of interconnection network. Based on the inputs that a neuron receives, a certain output is computed which is propagated to other neurons. A variety of models addresses activation status of neurons, determination of outputs and propagation of signals in the net with the basic goal to realize some kind of learning mechanism. The result computed by a neural network either appears explicitly as output or is given by the state of the neurons. Interesting in this context is the report HENRIQUES, SAFAYENI & FULLER (1987) which discusses quality of manual solutions to small TSP instances.
We discuss one approach for the Euclidean TSP in more detail (DURBIN & WILLSHAW (1987)). Here with each neuron a position in the plane is associated. In the beginning, the neurons are ordered along a circle. During the computation neurons are "stimulated" and approach a tour through the given set of points.

procedure neural_net

(1) Initialize a cycle of m neurons (e.g., $m = 3n$) in the area of the points defining the problem instance.

(2) As long as the stopping criterion is not satisfied perform the following steps.

(2.1) Choose a TSP point i at random.

(2.2) According to certain rules move the neuron whose position is closest to i and some of its neighbors in the neuron cycle towards i.

(3) Construct a tour from the final configuration of neurons.

end of neural_net

Figure 9.2 visualizes this neural net, showing that the cycle of neurons approaches a Hamiltonian tour. FRITZKE & WILKE (1991) give a further neural network algorithm for the TSP. A survey of different models is POTVIN (1993).
We think that computational results are not yet convincing. In our experiments, we found that the tours generated by the above algorithm are similar to nearest insertion tours with respect to structure and quality.

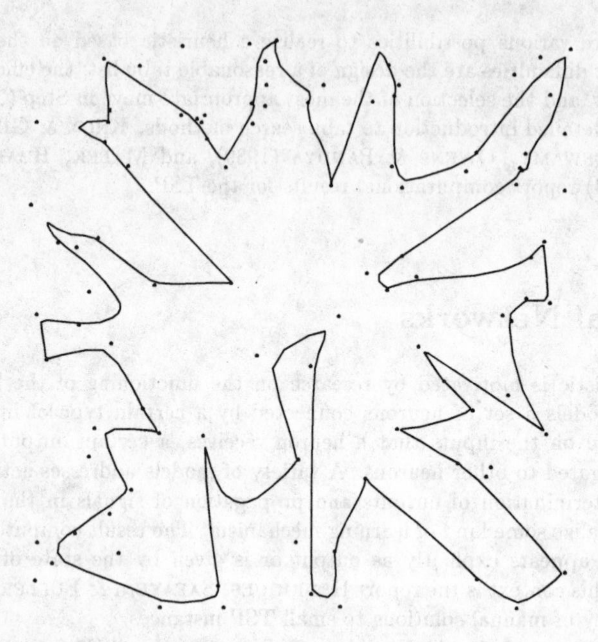

Figure 9.2 A neural net approaching a tour

For general reading on **neural networks** or **connectionism** see HOPFIELD & TANK (1985), RUMELHART, HINTON & MCCLELLAND (1986), and KEMKE (1988).

The **randomized improvement heuristics** or **stochastic search methods** discussed in this chapter certainly belong to the tool-kit of heuristic optimization for the TSP. To date, it is not clear whether there are variants that can compete with the purely deterministic approaches of Chapter 7 with respect to speed. If running time is not a major concern, then heuristics of this type can be successfully employed since they usually avoid bad local minima and have a chance to even find optimal solutions. We encourage the reader to consult the cited publications to form his/her own opinion and experiment with different variants. An attractive feature is, that these approaches are generally applicable to combinatorial optimization problems and other types of problems. They can be implemented with comparatively low effort and also with little knowledge about problem structure.

Chapter 10

Lower Bounds

When solving TSPs in practice, the main interest, of course, lies in computing good feasible tours. But, in addition, one would like to have some guarantee on the quality of the solutions found. Such guarantees can, except for weak a-priori guarantees for some heuristics, only be given if a lower bound for the length of a shortest possible tour is known.

In general, lower bounds are obtained by solving relaxations of the original problem in the sense that one optimizes over some set containing all feasible solutions of the original problem as a (proper) subset. Then (for a minimization problem) the optimal solution of the relaxed problem gives a valid lower bound for the value of the optimal solution of the original problem. Different relaxations provide different lower bounds. The main goal is to find relaxation problems for the TSP that can be solved efficiently and which are as tight as possible under this constraint.

For the purposes of this study, we are mainly interested in lower bounds which can be computed fast enough to only slightly increase the overall computation effort and running time. These lower bounds are not meant to give a very good estimate of the achievable optimum (say within a few percent) but to give indications on the quality of tours found by fast heuristics. This is usually sufficient for practitioners to get an impression of the performance of a heuristic on a particular problem.

But, to really evaluate a heuristic one should spend more time for computing better lower bounds. We will also address this question and indicate how CPU time can be saved.

10.1 Bounds from Linear Programming

Employing linear programming to determine lower bounds for the traveling salesman problem is not a major topic of this monograph. We will return to approaches for solving the TSP to optimality using linear programming bounds in Chapter 12. For the purposes of this chapter we need the theoretical framework provided by linear programming. The (combinatorial) bounds to be discussed in the following section can be compared with respect to related linear programming relaxations.

Consider the **traveling salesman polytope** $P_T = \text{conv}\{\chi^F \in \{0,1\}^{\binom{n}{2}} \mid \chi^F$ is the incidence vector of tour F in $K_n = (V_n, E_n)\}$, i.e., the convex hull of the incidence vectors of all tours in the complete graph. For a given vector c of edge lengths the optimal solution of the corresponding traveling salesman problem instance is obtained

by solving the linear programming problem $\min\{c^T x \mid x \in P_T\}$. The question of solving this problem will be addressed in Chapter 12.
Now let P be a polytope (or polyhedron) such that $P_T \subseteq P$. A solution of the problem $\min\{c^T x \mid x \in P\}$ yields a lower bound on the minimal tour length. Depending on how well P approximates P_T this lower bound may come close to the optimal tour length. Two relaxations are of particular interest for the following.

10.1.1 The 2-Matching Relaxation

A **perfect 2-matching** in a graph $G = (V, E)$ is a set of edges such that every node of V is incident to exactly two of these edges. Every tour is therefore a perfect 2-matching, but since also a collection of subtours is a perfect 2-matching we only have a relaxation. The following is an integer linear programming formulation of the 2-matching problem.

$$\begin{aligned} \min \sum_{ij \in E_n} & c_{ij} x_{ij} \\ x(\delta(i)) &= 2, && \text{for all } i \in V_n, \\ x_{ij} &\in \{0,1\}, && \text{for all } ij \in E_n. \end{aligned}$$

This problem can be solved in polynomial time (EDMONDS & JOHNSON (1973)). Implementation of this algorithm is nontrivial, its worst case running time is $O(n^3)$. An efficient implementation is discussed in PEKNY & MILLER (1994).
If we replace the requirement "$x_{ij} \in \{0,1\}$" by "$0 \le x_{ij} \le 1$" we obtain the **fractional 2-matching relaxation** of the traveling salesman problem.

10.1.2 The Subtour Elimination Relaxation

In the 2-matching problem only the degree constraints are taken into account. Short cycles are not forbidden. If we include conditions to eliminate such subtours we obtain the following integer linear program.

$$\begin{aligned} \min \sum_{ij \in E_n} & c_{ij} x_{ij} \\ \sum_{j=1}^{n} x_{ij} &= 2, && \text{for all } i \in V_n, \\ x(E(S)) &\le |S| - 1, && \text{for all } S \subseteq V_n, 2 \le |S| \le \lfloor \tfrac{n}{2} \rfloor, \\ x_{ij} &\in \{0,1\}, && \text{for all } ij \in E_n. \end{aligned}$$

Note that this formulation is equivalent to the one given in section 2.3. Feasible solutions of this problem are exactly the incidence vectors of tours. Therefore solving this problem is $\mathcal{NP}$-hard. Relaxing the integrality stipulations and observing that the constraints for the 2-element sets S yield upper bounds for the variables we obtain the **subtour**

elimination relaxation of the TSP.

$$\begin{aligned}
& \min \sum_{ij \in E_n} c_{ij} x_{ij} \\
& \sum_{j=1}^{n} x_{ij} = 2, && \text{for all } i \in V_n, \\
& x(E(S)) \le |S| - 1, && \text{for all } S \subseteq V_n, 2 \le |S| \le \lfloor \tfrac{n}{2} \rfloor, \\
& x_{ij} \ge 0, && \text{for all } ij \in E_n.
\end{aligned}$$

An equivalent formulation of this linear programming problem (GRÖTSCHEL & PADBERG (1985)) is the following.

$$\begin{aligned}
& \min \sum_{ij \in E_n} c_{ij} x_{ij} \\
& \sum_{j=1}^{n} x_{ij} = 2, && \text{for all } i \in V_n, \\
& x(\delta(W)) \ge 2, && \text{for all } W \subseteq V_n, 2 \le |W| \le \lfloor \tfrac{n}{2} \rfloor, \\
& x_{ij} \ge 0, && \text{for all } ij \in E_n.
\end{aligned}$$

Here subtour elimination constraints are given in their cut version. This corresponds to requiring that every (nonempty) cut in K_n contains at least two tour edges.
The subtour elimination bound can be determined in polynomial time using the ellipsoid method (GRÖTSCHEL, LOVÁSZ & SCHRIJVER (1988)) based on a polynomial time separation algorithm for subtour elimination constraints (PADBERG & GRÖTSCHEL (1985)). So far, no "nice" algorithm, i.e., an algorithm which does not explicitly need an LP solver as subroutine, for solving this problem in polynomial time is known. Some interesting properties of this relaxation are studied in BOYD & PULLEYBLANK (1990).

10.2 Simple Lower Bounds

We start the discussion of lower bounds for the TSP by considering several fairly simple bounds. These bounds are "combinatorial" in the sense that they are derived directly as obvious relaxations of the definition of tours.

10.2.1 The 1-Tree Bound

The **1-tree bound** for the TSP is based on the following observation. If we select some node of the problem, say node 1, then a Hamiltonian tour consists of a special spanning tree (namely a path) on the remaining $n-1$ nodes plus two edges connecting node 1 to this spanning tree. Hence we obtain a relaxation of the TSP if we take as feasible solutions arbitrary spanning trees on the node set $V_n \setminus \{1\}$ plus two additional edges

incident to node 1. Of course, in the case of nonnegative edge lengths, the weight of the minimum spanning tree alone also provides a (weaker) lower bound.
Figure 10.1 displays a 1-tree for problem pcb442. In this 1-tree the special node is the node in the lower right corner. The corresponding lower bound is 46858. Recall that a shortest possible tour has length 50778.

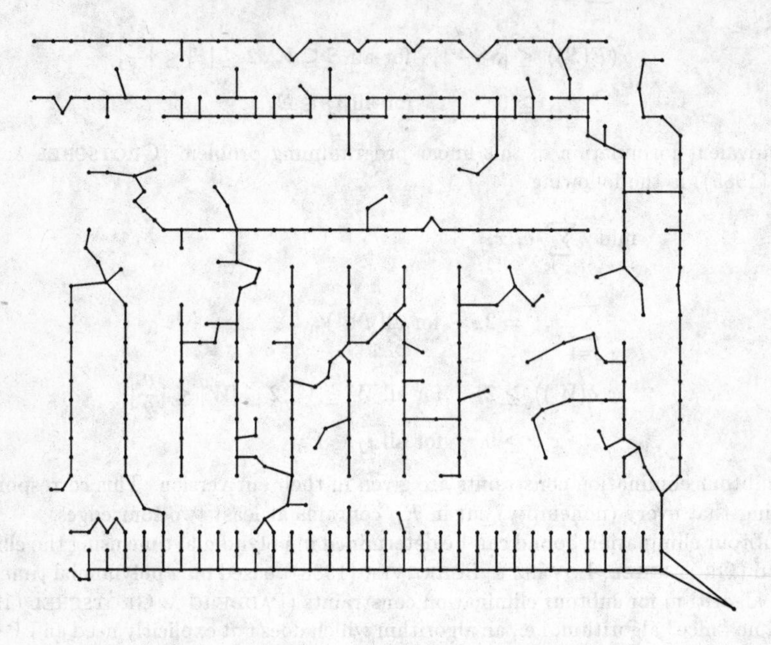

Figure 10.1 A 1-tree lower bound for pcb442

We use the following procedure to determine a 1-tree lower bound.

procedure simple_1tree

(1) Compute a minimum weight spanning tree T and let $c(T)$ be its weight.

(2) For every node i which is a leaf of this spanning tree compute the distance $d_2(i)$ to its second nearest neighbor (an edge to the nearest neighbor is already in T). This gives the lower bound $c(T) + d_2(i)$ on the minimal tour length.

(3) Take the best of the bounds computed in Step (2).

end of simple_1tree

Note, that we do not compute the best obtainable 1-tree. We just consider those nodes as special nodes which have degree 1 in the minimum spanning tree, and we take the best of these lower bounds. To compute the best 1-tree we have to compute n minimum spanning trees which is too time consuming.

In general, computing a minimum spanning tree in a complete graph on n nodes takes time $O(n^2)$, and finding the additional edge takes time $O(n)$ for every leaf. If we are not interested in computing several 1-trees we can compute a minimum spanning tree on the nodes $2, 3, \ldots, n$ in time $O(n^2)$ and add the two shortest edges incident to node 1 in time $O(n)$.

For geometric problems, we can do better by exploiting the Delaunay graph. Recall that the Delaunay graph is planar and therefore has $O(n)$ edges. Since the Delaunay graph contains a minimum spanning tree (for the complete graph) this spanning tree can be computed in the Delaunay graph in time $O(n \log n)$ using Kruskal's algorithm. Computing the various 1-trees for the leaves can be done with little additional time. For every node i we only have to compute the distances to nodes that are connected to i in the Delaunay graph by a path of length at most two. In the usual case these are only very few nodes.

The 1-tree computation constitutes a relaxation of the problem of finding a shortest Hamiltonian tour since we do not require every node to have degree 2. If a minimum 1-tree computed as above happens to satisfy this degree constraint, then it is an optimal tour. Unfortunately, this can never be expected for practical problems.

For other metrics we can employ the respective Delaunay graphs to speed up computations as well. With the help of the Delaunay graph we can determine the possible best 1-tree in time $O(n^2 \log n)$. Computing this 1-tree, however, turned out not to be worthwhile, because it only slightly (if at all) improves our simple 1-tree bound.

FRIEZE (1979) describes a tour construction heuristic based on 1-trees, that yields tours at most $2 - (k/n)$ times longer than an optimal tour in time $O(n^{3+k})$ for $1 \leq k \leq n-2$.

The 1-tree bound can be adapted in a fairly natural way to the asymmetric traveling salesman problem. For the ATSP, so-called **spanning 1-arborescences** are relaxations of directed Hamiltonian cycles. A 1-arborescence is an arc set with the property that every node has indegree at most 1 and a special node has indegree and outdegree equal to 1. The determination of minimum weight spanning 1-arborescences is more complicated than the determination of minimum spanning 1-trees, but can still be done very efficiently (FISCHETTI & TOTH (1993).

10.2.2 The 2-Neighbor Bound

In a tour, each node is connected to exactly two other nodes. If we have a tour in which each node is connected to its two nearest neighbors then this tour must be optimal.

Of course, this can only be achieved in very rare cases. In general, this observation leads to the derivation of a further simple lower bound. If we compute for all nodes the distances to their two nearest neighbors, sum up all these distances and divide by 2 then this number (rounded up) gives a lower bound on the minimal tour length. We call the subgraph of the complete graph which is obtained by taking for each node the two edges to its two nearest neighbors a **2-neighbor configuration**. In this case we keep multiple edges, so that such a configuration consists of $2n$ edges.

Figure 10.2 visualizes the 2-neighbor lower bound for problem pcb442. It provides the lower bound 47304.

The 2-neighbor bound can be computed trivially in time $O(n^2)$. For geometric instances we can do better by exploiting the Delaunay graph. Having computed the Delaunay graph in time $O(n \log n)$ we only have to compute for every node the distances to those nodes that are connected to it by a path of length at most two. The two nearest neighbors are among these nodes.

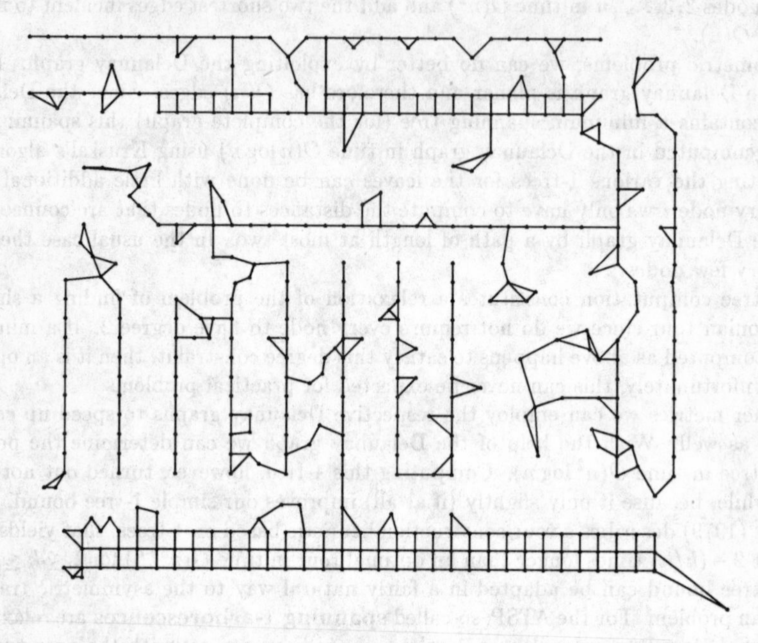

Figure 10.2 The 2-neighbor bound for pcb442

The quality of the bounds is, of course, highly problem dependent. The 2-neighbor configurations can give very poor bounds if there are clusters of points. Some such examples can be found in the computational results (e.g., d198, p654, u1060 or rl5934). A large gap between 1-tree and 2-neighbor bound might indicate that the problem is well suited for decomposition algorithms (described in Chapter 8) and also that the nearest neighbor candidate set is not sufficient for finding good tours.

10.2.3 The Assignment Relaxation

Another example of an relaxation that is often discussed in the context of TSP relaxations is the assignment relaxation. It is usually employed for asymmetric problem instances, but can also be formulated for the undirected problem.

An **assignment** for $V_n = \{1, 2, \ldots, n\}$ is a collection S of ordered pairs of the form $S = \{(i, n_i) \mid i = 1, 2 \ldots, n\}$ such that every node occurs exactly once as the second component of a pair.

A tour gives a particular assignment as follows: we choose one of the two possible orientations of the tour and assign to every node its successor in the tour according to the chosen orientation. Such assignments have the additional property that for every pair (i, n_i) we have $i \neq n_i$. The cost of the assignment is exactly the length of the tour.

Conversely, we can associate with every assignment the undirected graph given by the edges $\{i, n_i\}$, $i = 1, 2, \ldots, n$. This graph can have loops or multiple edges. If we restrict ourselves to assignments not containing pairs (i, n_i) with $i = n_i$ then the associated graph consists only of multiple edges and cycles.

Therefore, the assigment problem constitutes a relaxation of the TSP. It can be solved in time $O(n^3)$ with several algorithmic approaches, e.g. with the Hungarian method (see CARPANETO & TOTH (1983) for an efficient implementation).

If we do not allow loops, then the following is an integer linear programming formulation of the assignment relaxation for the TSP.

$$\min \sum_{ij \in E_n} c_{ij} x_{ij}$$

$$\begin{aligned} x(\delta(i)) &= 2, && \text{for all } i \in V_n, \\ x_{ij} &\in \{0, 1, 2\}, && \text{for all } ij \in E_n. \end{aligned}$$

The similarity to the 2-matching relaxation 10.1.1 is immediately seen. The important difference is that edge variables are allowed to have values greater than 1. Therefore, this relaxation is weaker than the 2-matching relaxation.

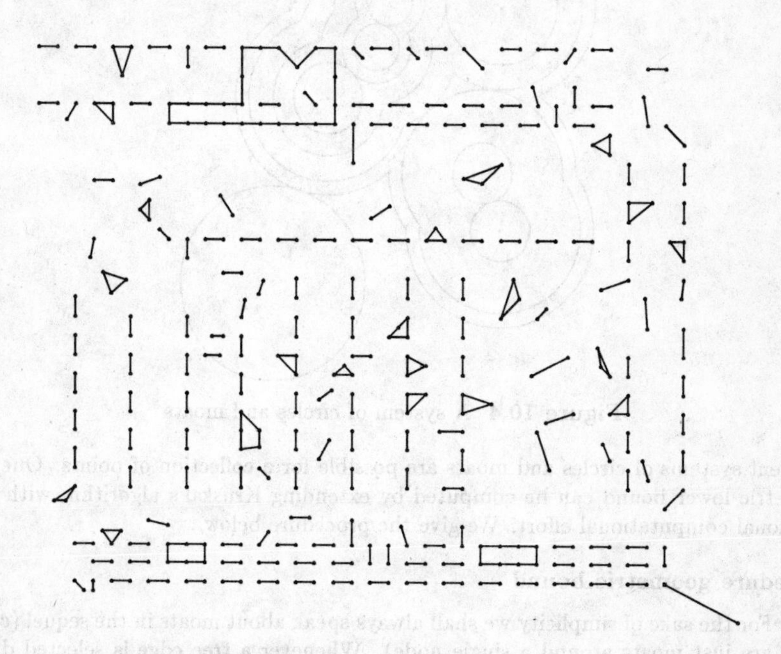

Figure 10.3 The assignment lower bound for pcb442

Figure 10.3 show the assignment lower bound for problem pcb442 with value 46830. The optimal assignment does not even come close to a tour. It contains many multiple edges and only short cycles.

10.2.4 Geometric Bounds

The idea of the geometric lower bound described in the sequel originates from techniques applied to the Euclidean matching problem (JÜNGER & PULLEYBLANK (1993)).
The geometric structure of Euclidean TSPs yields a very simple illustrative lower bound for the TSP. We compute a system of circles around nodes and moats around sets of circles and moats. This is done in such a way that circles and moats do not overlap each other. Moreover, there has to be always at least one node inside and outside of each circle and moat.
Each node has to be contained in a tour and every moat has to be crossed at least two times (since there are nodes inside and outside of each moat). Hence twice the sum of the radii of all circles and the width of all moats gives a lower bound on the minimal tour length.
Figure 10.4 gives an illustration of such a system consisting of 7 circles and 5 moats.

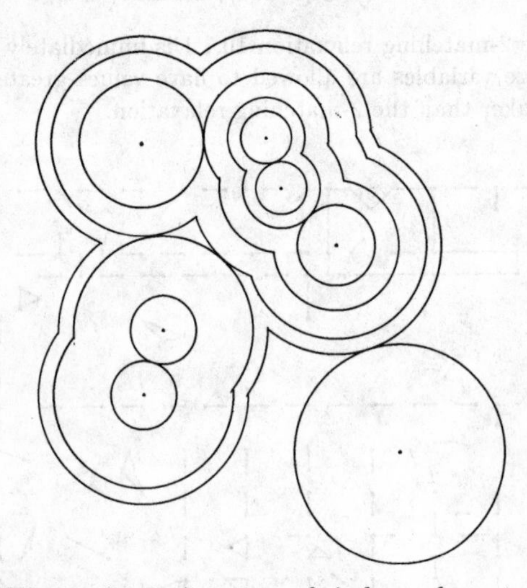

Figure 10.4 A system of circles and moats

Different systems of circles and moats are possible for a collection of points. One such geometric lower bound can be computed by extending Kruskal's algorithm with little additional computational effort. We give the procedure below.

procedure geometric_bound

(1) For the sake of simplicity we shall always speak about moats in the sequel (circles are just moats around a single node). Whenever a tree edge is selected during Kruskal's algorithm, it connects two components to form a new component. At this point a weight w will be assigned to the new component. This weight depends on the connecting edge and on the weights of the two participating components. Initially all components are just single nodes and their respective weights are 0. The lower bound lb is also set to 0.

(2) Let e be a tree edge with length c_e selected by Kruskal's algorithm. Let C_1 and C_2 be the two components to be connected to form the new component C. We set $lb := lb + 2 \cdot (c_e - w_{C_1} - w_{C_2})$ and $w_C := c_e/2$, where w_C, w_{C_1}, and w_{C_2} are the weights associated with the respective components.

(Note that the weights are not the moat widths. The widths of the moats around the components C_1 and C_2 used here implicitly are $\frac{1}{2}c_e - w_{C_1}$ and $\frac{1}{2}c_e - w_{C_2}$.)

(3) The value lb computed above gives a geometric lower bound based on circles and moats.

end of geometric_bound

The geometric lower bound as defined above can be obtained with (practically) no additional running time when implemented as in this procedure. Looking more closely at the way how this particular lower bound is computed, one realizes that it is exactly the sum of the length of a minimum spanning tree and the length of the final edge added by Kruskal's algorithm, i.e., the longest edge of the minimum spanning tree. Therefore, this specific computation is not restricted to geometric instances only. It applies to arbitrary TSPs.

The geometric bound seems to be rather weak. But we shall see below that the weak bounds obtained are only due to our simple scheme for determining the radii of the circles and the widths of the moats.

If we denote by z_i the radius of the circle around node i and by y_S the width of the moat around set S, $2 \leq |S| \leq n-1$, then the problem of finding the best bound can be formulated as a linear programming problem as follows.

$$\text{(CM)} \qquad \begin{aligned} \max \quad & 2\sum_{i=1}^{n} z_i + 2\sum_{S} y_S \\ & z_i + z_j + \sum_{\substack{S \\ i\in S, j\notin S}} y_S \leq c_{ij}, && \text{for all } ij \in E_n, \\ & z_i \geq 0, && \text{for all } i \in V_n, \\ & y_S \geq 0, && \text{for all } 2 \leq |S| \leq n-1. \end{aligned}$$

Dualizing this linear program we obtain

$$\text{(CM}_D\text{)} \qquad \begin{aligned} \min \quad & \sum_{ij\in E_n} c_{ij} x_{ij} \\ & \sum_{j=1}^{n} x_{ij} \geq 2, && \text{for all } i \in V_n, \\ & \sum_{\substack{S \\ i\in S, j\notin S}} x_{ij} \geq 2, && \text{for all } 2 \leq |S| \leq n-1, \\ & x_{ij} \geq 0, && \text{for all } ij \in E_n. \end{aligned}$$

Note that the first group of inequalities corresponds to a set S of cardinality 1 in the second system.

It is known that every vertex x^* of the polytope defining (CM_D) is given as the unique solution of a system of $\binom{n}{2}$ equations of the form

$$\begin{aligned} x_{ij} &= 0, \qquad \text{for all } ij \in F, \\ \sum_{i \in S, j \notin S} x_{ij} &= 2, \qquad \text{for all } S \in \mathcal{B} \subseteq 2^{V_n}, |S| \geq 1, \end{aligned}$$

where $\mathcal{B}$ is a nested family, i.e., for every $S_i, S_j \in \mathcal{B}$ we have either $S_i \subset S_j$, $S_j \subset S_i$, or $S_i \cap S_j = \emptyset$, and where $F \subseteq E_n, |F| = \binom{n}{2} - |\mathcal{B}|$ (CORNUEJOLS, FONLUPT & NADDEF (1985), BOYD & PULLEYBLANK (1990)). Due to this condition we obtain $|\mathcal{B}| \leq 2n-1$. In our application we have $c_{ij} > 0$ for all $ij \in E_n$ and we are looking for the minimizer of (CM_D) (i.e., the maximal circles and moats lower bound). Let x^* be the minimizer. It is easy to see that for every $i \in V_n$ we must have $\sum_{j=1}^{n} x_{ij} = 2$. Suppose this is not the case for some $i \in V_n$ and let S be the minimal element (with respect to inclusion) of $\mathcal{B}$ containing node i. Then there is an edge ik with $x^*_{ik} > 0$ and $k \in S$ (otherwise $S \notin \mathcal{B}$). We can set $x^*_{ik} = \max\{0, 2 - \sum_{j \neq k} x^*_{ij}\}$ without violating any condition and obtain a new solution having strictly less objective function value.
This proves that, for a nonnegative objective function, the best circles and moats bound is equivalent to the subtour relaxation bound. Therefore, as noted above, this bound can be determined in polynomial time making use of the ellipsoid method. It would be interesting to design efficient heuristics providing good systems of circles and moats.

10.2.5 Computations

We have evaluated the above bounds for our set of sample problems. Except for the case of the assignment bound all bounds were computed exactly. Because of the running time $O(n^3)$ of assignment algorithms for complete graphs, the assignment bound was only computed for the subgraph consisting of the Delaunay graph and the 10 nearest neighbor subgraph. We tested several cases and always found that the assignment computed for this subgraph had the same value as the minimum weight assignment computed for the complete graph.
Table 10.5 displays the qualities of the computed bounds relative to the best known upper bounds. I.e., if c_L is a lower bound, we define its **quality** as $100 \cdot (c_L - c_U)/c_U$ where c_U is the length of the best known tour as given in Table 3.1. Best qualities in each row are marked.
The table gives a clear picture. The 1-tree and the related geometric bound perform best, providing on the average bounds about 10% below the optimal objective function value. The 2-neighbor and the assignment bounds are significantly worse. In some cases, however, the 2-neighbor bound is better than the tree bounds. Therefore, since it is quickly computed, it is worthwhile to be also used as a fast lower bounding procedure accompanying the tree bounds.
We end this section with an impression of the necessary CPU times. Figure 10.6 shows the running times for minimum spanning tree, the simple 1-tree and the 2-neighbor configuration bounding procedures. Running times are given without the necessary preprocessing times for computing the Delaunay graph. The figure shows that very little additional time is needed and that CPU times are well predictable.

Problem	MST	1-tree	Geometric	2-neighbor	Assignment
d198	25.61	18.16*	17.08	37.31	32.78
lin318	9.94	9.21	8.78*	18.91	35.17
fl417	14.27	11.98	10.37*	30.66	35.50
pcb442	8.70	7.72	7.82	6.84*	7.78
u574	13.08	12.14*	12.25	17.03	20.97
p654	14.97	12.99	11.58*	29.11	32.14
rat783	7.73	7.44*	7.47	9.33	15.47
pr1002	13.46	12.82	12.66*	15.07	17.23
u1060	12.78	11.72*	11.96	16.49	18.15
pcb1173	9.63	9.20*	9.25	8.58	10.18
d1291	7.62	5.08*	5.11	16.98	19.83
rl1323	11.18	10.82	10.43*	18.93	23.32
fl1400	15.20	12.53*	12.90	31.01	39.40
u1432	4.57	4.38	4.39	3.39*	3.42
fl1577	12.62	12.00	10.67*	23.34	24.77
d1655	8.99	6.62*	6.64	12.15	13.73
vm1748	12.46	12.16*	12.20	14.40	17.01
rl1889	12.09	11.85	11.83*	18.71	22.87
u2152	4.16	4.00*	4.00*	7.80	12.10
pr2392	9.46	9.35	9.33*	10.66	15.49
pcb3038	7.55	7.42	7.42	6.69*	8.00
fl3795	12.18	10.33*	10.86	21.39	19.23
fnl4461	7.73	7.65*	7.66	6.83	9.97
rl5934	7.24	7.17	7.09*	13.64	17.76
Average	10.97	9.78	9.57	16.47	19.68

Table 10.5 Quality of lower bounding procedures

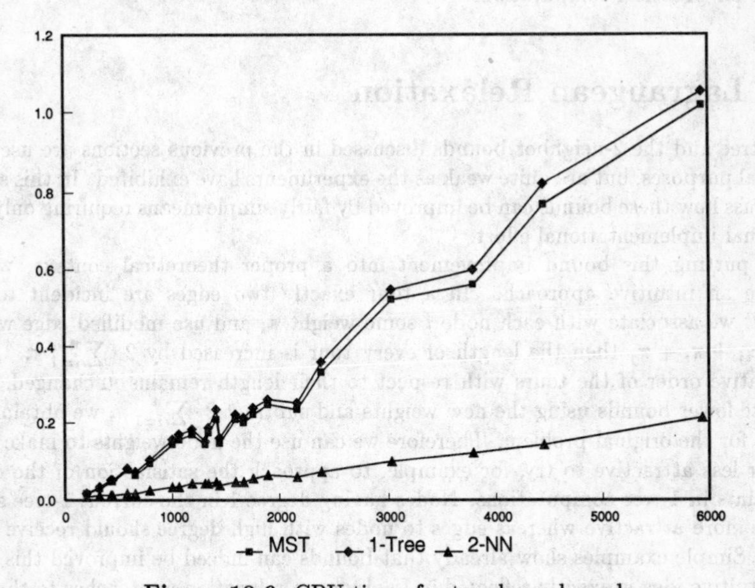

Figure 10.6 CPU times for tree lower bounds

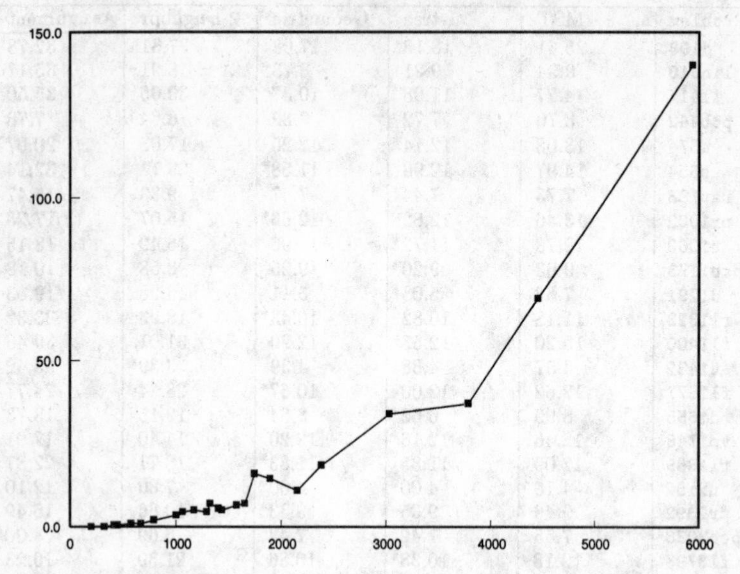

Figure 10.7 CPU times for assignment lower bounds

Figure 10.7 shows the running times for our implementation of the Hungarian method for solving assignment problems. Running time, even for the sparse graphs, are considerable. Therefore, the assignment bound is, not only because of its weakness, of no interest for practical computations.

10.3 Lagrangean Relaxation

The 1-tree and the 2-neighbor bounds discussed in the previous sections are useful for practical purposes, but are quite weak as the experiments have exhibited. In this section we discuss how these bounds can be improved by fairly simple means requiring only little additional implementational effort.

Before putting this bound improvement into a proper theoretical context, we will describe an intuitive approach. In a tour exactly two edges are incident to each node. If we associate with each node i some weight π_i and use modified edge weights $c'_{ij} = c_{ij} + \pi_i + \pi_j$ then the length of every tour is increased by $2 \cdot \sum_{i=1}^{n} \pi_i$. Hence the relative order of the tours with respect to their length remains unchanged. If we compute lower bounds using the new weights and subtract $2 \cdot \sum_{i=1}^{n} \pi_i$ we obtain lower bounds for the original problem. Therefore we can use the node weights to make nodes more or less attractive to try, for example, to approach the satisfaction of the degree constraints in 1-tree computations. Nodes having degree 1 in the current 1-tree should become more attractive whereas edges to nodes with high degree should receive larger weight. Simple examples show already that bounds can indeed be improved this way. This intuitive idea is exactly reflected in Lagrangean relaxation approaches to the TSP. We will first describe the general method.

10.3.1 The General Approach

One of the most common approaches for obtaining bounds on the optimal objective function value of an integer linear program is the method of **Lagrangean relaxation**. Let the following integer linear programming problem be given.

$$(\mathrm{P}) \qquad \begin{aligned} z^* := \quad & \min c^T x \\ & Ax = b \\ & Bx = d \\ & x \geq 0 \\ & x \text{ integer.} \end{aligned}$$

For any y the following integer linear programming problem provides a lower bound for the minimal value z^* of (P).

$$(\mathrm{D}_y) \qquad \begin{aligned} L(y) := \quad & \min c^T x + (d - Bx)^T y \\ & Ax = b \\ & x \geq 0 \\ & x \text{ integer.} \end{aligned}$$

The lower bound property can easily be seen since every feasible solution for (P) is also feasible for (D_y) with the same objective function value. Note that, since y is not restricted in sign, we could as well use the objective function $c^T x + (Bx - d)^T y$. The vector y is called vector of **Lagrange multipliers**. The best such lower bound is then given by solving the so-called **Lagrangean dual problem**

$$(\mathrm{LD}) \qquad u^* := \max_y L(y)$$

i.e., by finding the maximum of the function L.
The function L is piecewise linear and concave (and hence nondifferentiable). A suitable method for maximizing L is **subgradient maximization**. A vector d is called **subgradient** of L at x, if $d^T(y - x) \geq L(y) - L(x)$ for all y.
Suppose x^* is the minimizer of D_{y^*} for some vector of Lagrange multipliers y^*. Then $u^* = d - Bx^*$ is a subgradient of L at y^* as is seen from

$$\begin{aligned} L(y^* + h) - L(y^*) \; & \leq c^T x^* + (d - Bx^*)^T (y^* + h) - c^T x^* - (d - Bx^*)^T y^* \\ & = (d - Bx^*)^T h \\ & = h^T u^*, \text{ for all } h. \end{aligned}$$

Hence solving problem D_y for some y, at the same time provides a subgradient of L at y.
The **subgradient method** is an iterative method which at a given point y^k computes the next iterate y^{k+1} by

$$y^{k+1} = y^k + \lambda_k d^k,$$

where d^k is a subgradient of L at y^k and λ_k is a suitable step length.
If L is bounded from above and if the step lengths satisfy both $\lim_{k\to\infty} \lambda_k = 0$ and $\sum_{k=0}^{\infty} \lambda_k = \infty$, then the method converges to the maximum of L (POLYAK (1978)).
It turned out in practice that the step length formula above leads to very slow convergence. So the requirement $\sum_{k=0}^{\infty} \lambda_k = \infty$ is usually dropped and there are several formulas for computing the step length leading to satisfactory convergence in practical applications. A widely used formula is e.g.,

$$\lambda_k = \alpha \cdot \frac{U - L(y^k)}{||d^k||},$$

where U is an upper bound on L, and $0 < \alpha < 2$ is some constant which is periodically decreased. The method is stopped as soon as there is no more significant increase in the bound.
Clearly, the quality of the bound provided by the Lagrangean dual depends on the choice of the constraint set $Bx = d$ to be relaxed. This also influences the complexity of the Lagrangean subproblems to be solved for each multiplier set y.
Consider the following series of problems.

$$(\mathrm{P}) \qquad z^* := \min_x \{c^T x \mid Ax = b, Bx = d, x \geq 0, x \text{ integer }\}$$

$$(\mathrm{P_{LP}}) \qquad z^*_{LP} := \min_x \{c^T x \mid Ax = b, Bx = d, x \geq 0\}$$

$$z^*_D := \max_{u,v} \{d^T u + b^T v \mid u^T B + v^T A \leq c^T\}$$

$$u^*_D := \max_u \{d^T u + \max_v \{b^T v \mid v^T A \leq c^T - u^T B\}\}$$

$$u^*_{LP} := \max_u \{d^T u + \min_x \{(c^T - u^T B)x \mid Ax = b, x \geq 0\}\}$$

$$(\mathrm{LD}) \qquad u^* := \max_u \{d^T u + \min_x \{(c^T - u^T B)x \mid Ax = b, x \geq 0, x \text{ integer }\}\}$$

In general we have $z^* \geq z^*_{LP} = z^*_D = u^*_D = u^*_{LP} \geq u^*$.
In the special case where $\{x \mid Ax = b, x \geq 0\}$ is an integer polyhedron (i.e., has only integer vertices), we have $u^* = u^*_{LP} = z^*_{LP}$. In this case, the value of the Lagrangean dual is equal to the value of the linear programming relaxation ($\mathrm{P_{LP}}$) of the integer linear program (P). Moreover, as noted in SCHRIJVER (1986), if we can solve $\min\{(c^T - u^T B)x \mid Ax = b, x \geq 0\}$ in polynomial time for each u, then we can also solve (LD) in polynomial time and we can directly apply LP techniques on ($\mathrm{P_{LP}}$) to compute u^*. But, even though (LD) might be solvable in polynomial time, this fact might not be exploitable in practice, especially when larger problems have to be solved. Therefore, for each choice of Lagrangean relaxation one has to think about the best way for computing or approximating u^*.

For convenience we have used equation systems for both the relaxed and the maintained constraints. Treating inequality systems does not cause any problems. If inequalities ("$\geq$") are relaxed then their corresponding Lagrange multipliers must be nonnegative. We will now show that 1-tree and 2-neighbor relaxations are particular applications of the general method for the TSP.

10.3.2 Lagrangean Relaxation with 1-Trees

In section 2.3 we have given an integer linear programming formulation of the TSP. For the purposes of this chapter we write this formulation in a different, but equivalent form.

$$
\begin{aligned}
c^* := \quad & \min \sum_{ij \in E_n} c_{ij} x_{ij} \\
& \sum_{j=1}^{n} x_{ij} = 2, && \text{for } i \in V_n \setminus \{1\}, \\
& \sum_{j=1}^{n} x_{1j} = 2, \\
& \sum_{ij \in E_n} x_{ij} = n, \\
& x(C) \leq |C| - 1, && \text{for all cycles } C \text{ in } \{2,3,\ldots,n\}, \\
& x_{ij} \in \{0,1\}, && \text{for all } ij \in E_n.
\end{aligned}
$$

If we now relax the first system of equations and associate multipliers π_i with the nodes, we obtain the following relaxation. For convenience we also define π_1 and set it to 0.

$$
\begin{aligned}
L(\pi) := \quad & \min -2 \sum_{i=1}^{n} \pi_i + \sum_{ij \in E_n} (c_{ij} + \pi_i + \pi_j) x_{ij} \\
& \sum_{j=1}^{n} x_{1j} = 2, \\
& \sum_{ij \in E_n} x_{ij} = n, \\
& x(C) \leq |C| - 1, && \text{for all cycles } C \text{ in } \{2,3,\ldots,n\}, \\
& x_{ij} \in \{0,1\}, && \text{for all } ij \in E_n.
\end{aligned}
$$

It is known that the condition $x_{ij} \in \{0,1\}$, for all $ij \in E_n$, is not necessary in this formulation because the system of linear equations and inequalities describes an integral polyhedron whose vertices are exactly the incidence vectors of 1-trees. Therefore, due to the observations above, the value of the Lagrangean dual based on 1-trees is equivalent to the lower bound provided by the subtour elimination relaxation defined in 10.1.2. Since feasible solutions of the integer programm are exactly 1-trees with special node 1, the relaxation can be rewritten as follows.

$$L(\pi) := \quad \min \sum_{ij \in E_n} c_{ij}x_{ij} + \sum_{i=1}^{n} \pi_i (\sum_{j=1}^{n} x_{ij} - 2)$$

$$x \text{ is incidence vector of a 1-tree .}$$

Determining $L(\pi)$ for a given π amounts to computing a 1-tree with respect to the modified edge weights $c_{ij}+\pi_i+\pi_j$ and subtracting $2\sum_{i=1}^{n} \pi_i$. According to the preceding section, the minimum 1-tree readily supplies a subgradient as follows. Let δ_i be the degree of node i in the minimum 1-tree. Then the vector $(\delta_1 - 2, \delta_2 - 2, \ldots, \delta_n - 2)$ is a subgradient of L at π.
HELD & KARP (1970,1971) describe the first algorithm for finding the maximum of L, and we therefore call the best 1-tree bound also **Held-Karp bound.** There are several variations concerning the choice of initial step sizes and update of step sizes (HELBIG-HANSEN & KRARUP (1974), SMITH & THOMPSON (1977), VOLGENANT & JONKER (1982), and BALAS & TOTH (1985)). Based on these references and on own experiments we used the following implementation.

procedure 1tree_bound

(1) Let τ be the initial step length, λ a decrement factor for the step length, and m the number of iterations.

(2) Set $t^1 = \tau$, $\pi_i^1 = 0$ for every node i, and $k = 1$.

(3) As long as $k \leq m$ perform the following steps.

(3.1) Compute a minimum spanning tree with respect to the edge weights $c_{ij} + \pi_i + \pi_j$.

(3.2) Compute the best 1-tree obtainable from this spanning tree (section 10.2).

(3.3) Define the vector d^k by $d_i^k = \delta_i - 2$, where δ_i is the degree of node i in the 1-tree computed in Step (3.2).

(3.4) For every node i set

$$\pi_i^{k+1} = \pi_i^k + t^k(0.7d_i^k + 0.3d_i^{k-1}).$$

(3.5) Set $t^{k+1} = \lambda t^k$ and increment k by 1.

(4) Return the best bound computed.

end of 1tree_bound

Differences to straightforward realizations are, that the direction of the subgradient step is a convex combination of the current and the preceding subgradient, that the direction vector is not normalized, and that the special node for the 1-tree computations is not fixed. In theory, the same optimal value of the Lagrange dual is attained whatever node is fixed. Actually, it is even incorrect to have varying nodes because the underlying optimization problem changes. But practical experiments have shown that better bounds

are obtained this way and that it pays off to spend the additional running time for computing various 1-trees.
Some authors propose to update the multipliers according to the formula

$$\pi_i^{k+1} = \pi_i^k + t^k(U - L(\pi^k))\frac{d_i^k}{||d^k||}$$

where U is an estimate for the optimal solution and $L(\pi^k)$ is the currently computed lower bound. We found that convergence is faster with this formula, but bounds are inferior.
A general guideline for the performance is the following. If λ is close to 1 (say 0.98–0.995) convergence is slow, but usually better bounds are reached. For values of λ between 0.95 and 0.97 faster convergence is achieved yielding reasonable bounds. Smaller values of λ lead to considerably inferior bounds. Instead of fixing the number of iterations one can stop, if no significant progress is observed any more.
Since no line search is performed it is not guaranteed that each iteration step improves the bound. In fact, the behaviour shown in Figure 10.8 can be observed. This figure displays the development of the bounds during application of the above subgradient method to problem pcb442 for 150 iterations.

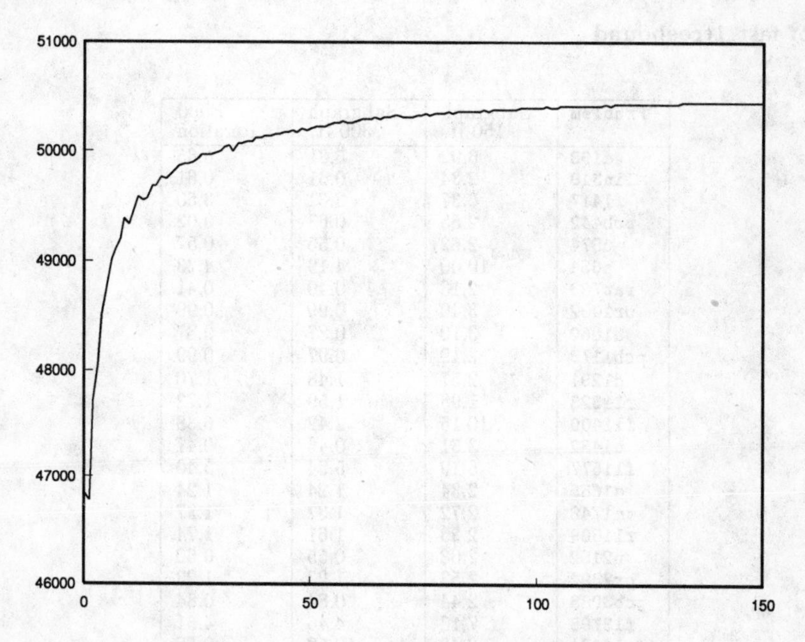

Figure 10.8 A run of the subgradient algorithm

The best bound obtained is 50459. In general, the evolution of the lower bounds does not have to be as smooth as in this example. Depending on the problem instance the

use of non-fixed special nodes can have the consequence that the bounds alternate even at the end of the procedure. But, our emphasis does not lay on nice convergence but on good bounds. Hence this is tolerable.
Concerning running time we are in a bad situation. Since edge weights are arbitrary we can compute the best tree in Step (3.1) only in time $O(n^2)$ and the best 1-tree in Step (3.2) in time $O(kn)$ where k is the number of leaves of the spanning tree.
To speed up the procedure we compute trees in sparse subgraphs. This has the consequence that the computed bound may not be valid for the true problem. Only if the subgraph contains an optimal tour for the original problem, the bound is valid. Of course, this cannot be verified. We therefore proceeded as follows for our sample problems.

procedure fast_1treebound

(1) Construct the subgraph consisting of the 10 nearest neighbor edges and the edges of the Delaunay graph.

(2) Perform the subgradient algorithm only in this graph to compute an approximate lower bound.

(3) Compute a minimum 1-tree in the complete graph using the multipliers of the final iteration in (2).

end of fast_1treebound

Problem	Subgraph 150 It.	Subgraph 300 It.	Final iteration
d198	6.92	5.31	5.65
lin318	2.34	0.61	0.61
fl417	6.37	3.31	3.50
pcb442	2.55	0.63	0.63
u574	2.62	0.55	0.57
p654	10.00	4.19	4.23
rat783	2.13	0.40	0.41
pr1002	3.40	0.99	0.99
u1060	3.10	0.87	0.87
pcb1173	2.19	0.97	0.99
d1291	2.37	1.48	1.70
rl1323	1.98	1.59	1.72
fl1400	10.15	2.42	6.38
u1432	3.31	0.46	0.47
fl1577	6.19	5.34	5.40
d1655	2.34	1.24	1.24
vm1748	2.72	1.37	1.37
rl1889	2.29	1.61	1.74
u2152	2.08	0.55	0.60
pr2392	2.52	1.23	1.23
pcb3038	2.44	0.84	0.84
fl3795	7.13	4.46	4.61
fnl4461	2.49	0.58	0.58
rl5934	1.63	1.20	1.23
Average	3.80	1.76	1.98

Table 10.9 Results of Lagrangean relaxation based on 1-trees

Running time is now $O(n \log n) + O(kn)$ for each iteration in Step (2) where k is the number of leaves of the spanning tree and $O(n^2)$ for the final step. Due to this final optimization a valid lower bound is determined.
Table 10.9 documents a sample run with this fast bounding procedure where we have set $\lambda = 0.98$ and $m = 300$. As initial step length we choose $10 \cdot (U - T)/n$ where U is the tour length shown in Table 6.17 for the Christofides starting tour and T is the 1-tree bound listed in Table 10.5.
We give the bounds obtained from the subgraph optimization after 150 and after 300 iterations and the bound obtained in the final 1-tree computation. The table verifies that our approach is indeed reasonable. The final iteration changes the bound only marginally. In practical applications we can safely omit the final step and assume that the bound determined in the first phase is correct.
Respective CPU times are given in Figure 10.10. They do not increase smoothly because the running time of Kruskal's algorithm depends on the distribution of edge lengths for the respective subgraphs. This distribution influences the number of edges that are checked for entering the spanning tree. A uniform distribution usually leads to earlier termination.

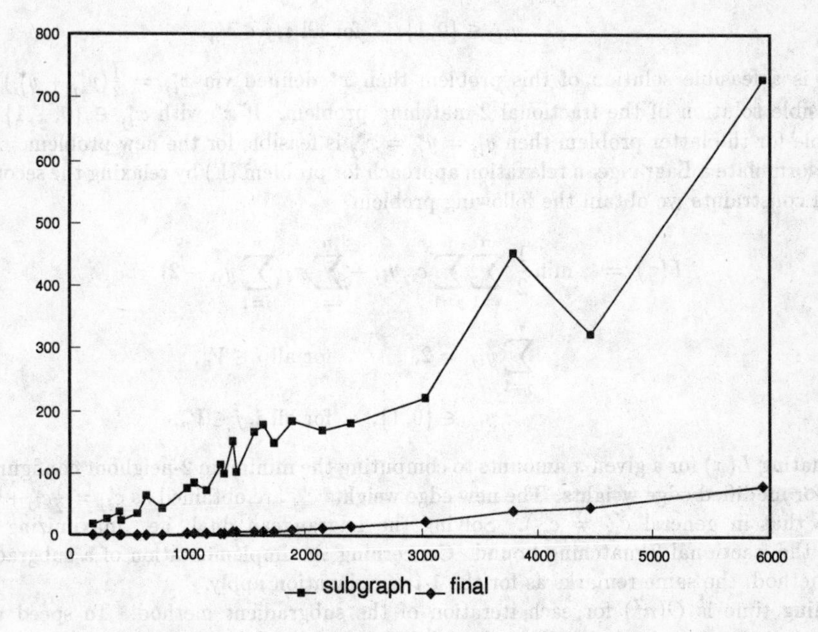

Figure 10.10 CPU times for Lagrangean relaxation (1-tree)

Methods for finding the optimum of the Lagrangean dual based on 1-trees are not limited to subgradient approaches. One further iterative method is, e.g., dual ascent (MALIK & FISHER (1990)). As note above, a completely different way is to directly solve the linear program corresponding to the subtour elimination relaxation.

10.3.3 Lagrangean Relaxation with 2-Neighbor Configurations

We will now show that the 2-neighbor configurations are related to the fractional 2-matching relaxation. It is known (BALINSKI (1970)) that the polytope defining the fractional 2-matching problem has only vertices whose components have values in $\{0, \frac{1}{2}, 1\}$. To treat 2-neighbor configurations we switch to (directed) arcs. We introduce variables y_{ij} with the interpretation that $y_{ij} = 1$ if the arc from i to j is selected, and $y_{ij} = 0$ otherwise. We set also $c_{ij} = c_{ji}$ and for notational convenience, we use variables y_{ii} in the formulae (which then have to be ignored).
Consider the following integer programming problem

$$\text{(F)} \qquad \begin{aligned} \min \frac{1}{2} \sum_{i=1}^{n} \sum_{j=1}^{n} c_{ij} y_{ij} & \\ \sum_{j=1}^{n} y_{ij} &= 2, && \text{for all } i \in V_n, \\ \sum_{j=1}^{n} y_{ji} &= 2, && \text{for all } i \in V_n, \\ y_{ij} &\in \{0,1\}, && \text{for all } i, j \in V_n. \end{aligned}$$

If y^* is a feasible solution of this problem then x^* defined via $x^*_{ij} = \frac{1}{2}(y^*_{ij} + y^*_{ji})$ is a feasible solution of the fractional 2-matching problem. If x^* with $x^*_{ij} \in \{0, \frac{1}{2}, 1\}$ is feasible for the latter problem then $y^*_{ij} = y^*_{ji} = x^*_{ij}$ is feasible for the new problem.
If we formulate a Lagrangean relaxation approach for problem (F) by relaxing the second set of constraints we obtain the following problem

$$\begin{aligned} L(\pi) := \quad & \min \frac{1}{2} \sum_{i=1}^{n} \sum_{j=1}^{n} c_{ij} y_{ij} + \sum_{i=1}^{n} \pi_i \Big(\sum_{j=1}^{n} y_{ji} - 2\Big) \\ & \sum_{j=1}^{n} y_{ij} = 2, && \text{for all } i \in V_n, \\ & y_{ij} \in \{0,1\}, && \text{for all } i, j \in V_n. \end{aligned}$$

Evaluating $L(\pi)$ for a given π amounts to computing the minimum 2-neighbor configuration for modified edge weights. The new edge weights c'_{ij} are obtained as $c'_{ij} = \frac{1}{2} c_{ij} + \pi_j$ (note that in general $c'_{ij} \neq c'_{ji}$). Solving the Lagrangean dual, i.e., maximizing L gives the fractional 2-matching bound. Concerning the implementation of a subgradient method, the same remarks as for the 1-tree relaxation apply.
Running time is $O(n^2)$ for each iteration of the subgradient method. To speed up computations,we optimize in subgraphs only. If the subgraph has m edges each iteration takes time $O(m)$ and the final iteration takes time $O(n^2)$ to yield a valid lower bound. We ran experiments using the 10 nearest neighbor subgraph augmented by the edges of the Delaunay graph. In this case each iteration runs in time $O(n)$. Table 10.11 shows the results. Again, the final step alters the bounds only slightly and the approach is verified to be reasonable.

Problem	Subgraph 150 It.	Subgraph 300 It.	Final iteration
lin318	7.50	7.43	7.43
fl417	22.68	22.50	24.37
pcb442	1.63	1.34	1.34
u574	7.28	7.18	7.18
p654	18.39	17.53	17.71
rat783	2.83	2.70	2.70
pr1002	7.09	7.01	7.02
u1060	6.74	6.51	6.51
pcb1173	2.35	2.27	2.27
d1291	7.60	7.54	7.54
rl1323	9.37	9.27	9.28
fl1400	16.72	15.52	15.54
u1432	1.45	0.88	0.88
fl1577	19.48	19.28	19.28
d1655	5.37	5.24	5.24
vm1748	4.58	4.46	4.49
rl1889	9.23	9.11	9.12
u2152	4.45	4.22	4.22
pr2392	4.99	4.87	4.87
pcb3038	1.77	1.67	1.67
fl3795	16.09	15.49	15.49
fnl4461	1.90	1.82	1.82
rl5934	5.92	5.86	5.86
Average	8.79	8.54	8.63

Table 10.11 Results of Lagrangean relaxation based on 2-neighbors

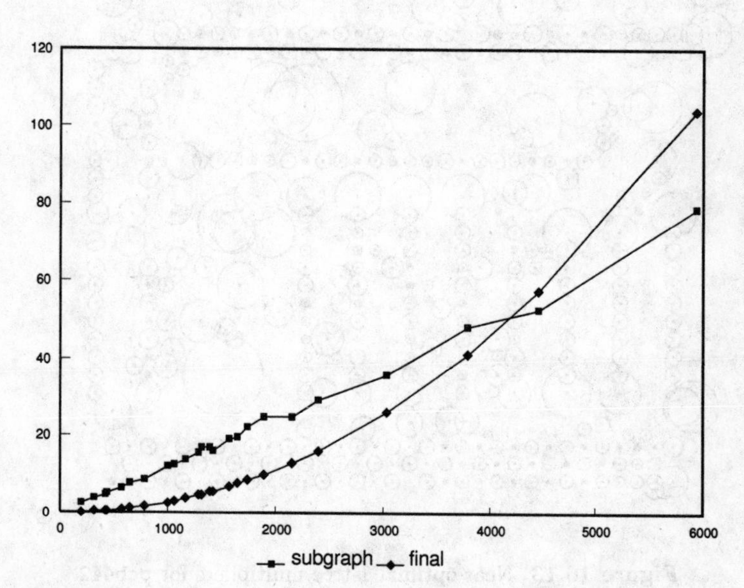

Figure 10.12 CPU times for Lagrangean relaxation (2-neighbor)

CPU time statistics show that time per iteration is considerably less for the 2-neighbor relaxation than for the 1-tree relaxation. Therefore, if one observes the development of the bounds over time, for some problems during a certain first time period the 2-neighbor bound will we above the 1-tree bound. In the end, according to the theoretical results, the 1-tree bound will be better.
Lagrangean relaxations based on perfect 2-matchings are considered in SMITH, MEYER & THOMPSON (1990).

10.3.4 Multiplier Heuristics

Every assignment of node multipliers can be used to determine a lower bound using 1-trees or 2-neighbor configurations. The simple lower bounds implicitly use zero multipliers, the subgradient method adapts the multipliers starting with zero multipliers. In this section we briefly address the question of using a heuristic for guessing reasonable multipliers.
To get an impression on how the π_i values look like we show in Figure 10.13 near optimal multipliers for the 1-tree relaxation for problem pcb442 giving the lower bound 50490. Figure 10.14 displays multipliers for the 2-neighbor relaxation giving a lower bound of 50099.

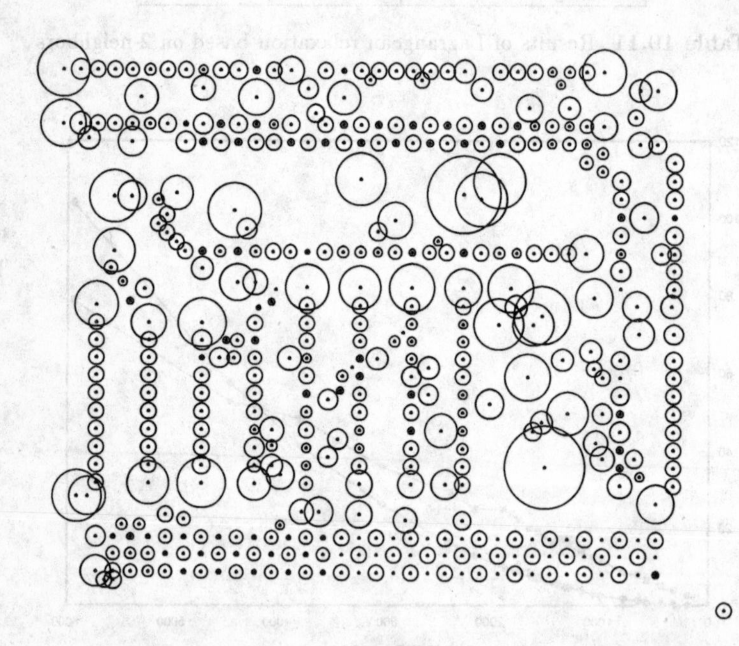

Figure 10.13 Near optimal 1-tree multipliers for pcb442

Note that one can add the same constant to all multipliers without affecting the lower bounds obtained by the relaxation. We have therefore subtracted the maximum π_i

from all multipliers. Now all values are nonpositive and the largest one is zero. Radii of circles in the figures are the complemented π_i values.
The only hint for estimating reasonable multipliers is that apparently isolated nodes have large negative multipliers whereas the other nodes have multipliers related to their nearest neighbor distance. This corresponds to the intuitive explanation that isolated nodes have to be made more attractive since otherwise they would become leaves of the 1-tree.

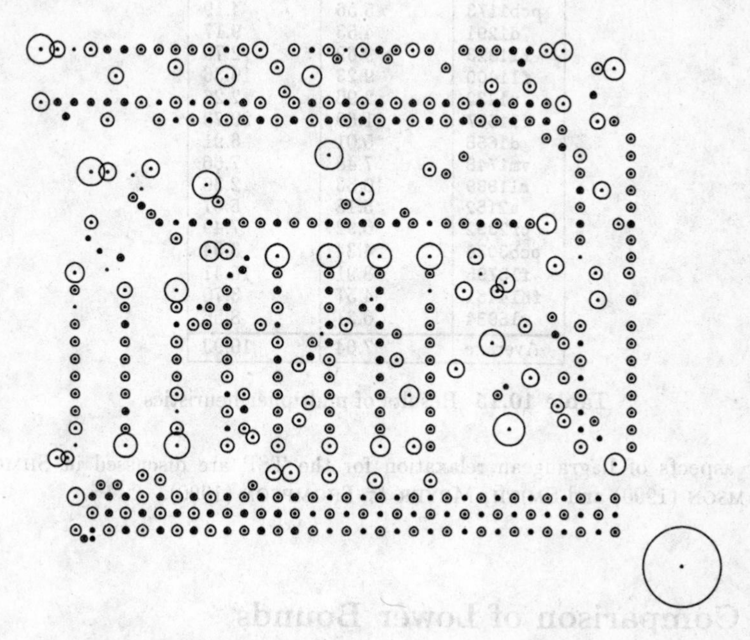

Figure 10.14 Near optimal 2-neighbor multipliers for **pcb442**

We have implemented three variants of multiplier heuristics which also run fast. Let $d_1(i)$ be the distance to the nearest neighbor of i and $d_2(i)$ be the distance to the second nearest neighbor. The variants are

(1) Set $\pi_i = -0.5d_1(i)$.

(2) Set $\pi_i = -0.5d_2(i)$.

(3) Set $\pi_i = -0.25(d_1(i) + d_2(i))$.

We have seen in preceding chapters that nearest neighbor distances for Euclidean problems can be computed very efficiently. For the general problem we need time $O(n^2)$ to compute these multipliers. In any case, eve if multipliers can be computed fast, we have to spend time $O(n^2)$ for proving the validity of the bounds.
Table 10.15 displays the results obtained with Variant 2 for the 1-tree and the 2-neighbor relaxation which are significantly better than the simple bounds of Table 10.5. The other two variants performed worse, so their results are not documented.

Problem	1-tree	2-nn
d198	15.96	26.76
lin318	6.83	12.32
fl417	12.78	26.52
pcb442	4.53	2.51
u574	9.11	10.02
p654	10.70	19.78
rat783	5.28	5.30
pr1002	9.13	9.36
u1060	8.28	9.24
pcb1173	5.56	4.19
d1291	4.53	9.17
rl1323	9.69	12.79
fl1400	9.23	19.36
u1432	2.96	2.26
fl1577	11.69	20.73
d1655	5.01	6.91
vm1748	7.48	7.66
rl1889	10.33	12.36
u2152	3.18	5.57
pr2392	6.92	7.49
pcb3038	4.34	3.34
fl3795	8.91	16.41
fnl4461	4.57	3.70
rl5934	6.31	8.56
Average	7.64	10.93

Table 10.15 Results of multiplier heuristics

Further aspects of Lagrangean relaxation for the TSP are discussed in SHMOYS & WILLIAMSON (1990) and SMITH, MEYER & THOMPSON (1990).

10.4 Comparison of Lower Bounds

We have seen that the optimal objective function value c_T of the Lagrangean dual based on 1-trees is the subtour elimination bound. The Lagrangean dual based on 2-neighbor configurations gives the fractional 2-matching bound c_N. Since the latter is itself a relaxation of the subtour relaxation we get that $c_N \leq c_T$. Normally, we have $c_N < c_T$. For practical problems the difference between the two bounds will be considerable.

To examine how well our subgradient algorithm approximates the optima c_N and c_T we have also computed the exact values using LP techniques. We used an branch and cut code for solving TSPs optimally (JÜNGER, REINELT & THIENEL (1993)) to compute these bounds for all of our sample problems. Table 10.16 displays the results and shows that in some cases (e.g., d198, p654, or rl1323) the 1-tree bound computed in Table 10.9 misses the optimal 1-tree bound by some percent. Having a look at such problems, one realizes that they are built of clusters of points. For such instances, our subgradient method has problems in approaching the best bound. We can only overcome this by enlarging the decrement factor and hence coming closer to the theoretically required formula for obtaining convergence to the optimum of the Lagrangean dual. For example, if we use $\lambda = 0.995$, then for problem d198 we obtain the lower bound 14769 after 800 iterations. If the points are more or less uniformly distributed we have no difficulties in finding rather close approximations to the best bound.

Problem	Fractional 2-Matching	Subtour elimination
d198	11793	15712
lin318	38964	41889
fl417	8961	11790
pcb442	50104	50500
u574	34256	36714
p654	28584	34596
rat783	8568	8773
pr1002	240878	256766
u1060	209529	222651
pcb1173	55600	56351
d1291	46971	50209
rl1323	245156	265815
fl1400	16783	19783
u1432	151676	152535
fl1577	17871	22134
f1655	58876	61544
vm1748	321552	332061
rl1889	287698	311705
u2152	61464	63859
pr2392	359620	373490
pcb3038	135391	136588
fl3795	24289	28478
fnl4461	179252	181570
rl5934	521629	548471

Table 10.16 Exact values of relaxations

Finally, in Table 10.17, we give the average qualities of all relaxations discussed in this chapter taking only those 19 sample instances into account where true optimal solutions are known.

Name of heuristic	Average deviation from optimum
Subtour Elimination	0.78
Lagrange 1-Tree bound (final step)	1.54
Multiplier heuristic (1-Tree)	7.58
Fractional 2-Matching	7.70
Lagrange 2-NN bound (final step)	7.72
Geometric bound	9.70
Simple 1-tree	9.93
Multiplier heuristic (2-NN)	10.09
Minimum spanning tree	11.15
Simple 2-neighbors	15.69
Assignment	18.90

Table 10.17 Comparison with optimal solutions

The table shows that the fractional 2-matching bound is about 7% below the subtour elimination bound. Our fast schemes to approximate the subtour bound and the fractional 2-matching bound perform fairly well on the average. Whereas the fractional 2-matching bound is almost met by our heuristic, the subtour bound is missed by about 1%. As we pointed out, we can improve the approximation by tuning the parameters of the subgradient algorithm.

Multiplier heuristics improve the simple bounds considerably. If Delaunay graphs are available, then at least the simple bound computations should be performed in any case.

The subgradient method used in this chapter is not the only way for attacking nondifferentiable optimization problems. A more elaborate approach is the so-called **bundle method** (KIWIEL (1989)). It is also based on subgradients, but in every iteration the new direction is computed as a convex combination of several (10–20) previous subgradients. Moreover, line searches are performed. In this sense, our approach is a simple version of the bundle method keeping only a "bundle" of two subgradients (which are combined in a fixed way) and not performing line searches.

SCHRAMM (1989) considers an extension of this principle which combines the bundle approach with trust-region methods. Whereas, in general, this algorithm outperforms pure subgradient methods this is not the case for the 1-tree relaxation. Here performance is similar.

Several further relaxations are available for the TSP. Among them are ***n*-path relaxation** or so-called **additive bounding** procedures. For information on further approaches see HOUCK, PICARD, QUEYRANNE & VEMUGANTI (1980), BALAS & TOTH (1985), MACULAN & SALLES (1989). CARPANETO, FISCHETTI & TOTH (1989), and FISCHETTI & TOTH (1992).

Chapter 11

A Case Study: TSPs in Printed Circuit Board Production

The wide range of applicability of the TSP covers in particular some problems arising in design and production of printed circuit boards and very large scale integrated circuits. In this chapter we will discuss two such applications in depth and show how the methods developed in the previous chapters can be successfully employed for practical problem solving. Results for real world data sets are presented.

The two applications we consider occur in the process of manufacturing printed circuit boards. Even more combinatorial optimization problems have to be treated in the design phase which is not discussed here. One problem (**plotting of masks**) is to be solved at an early stage of the production process while the other problem (**drilling of boards**) constitutes a final step.

Since the latter problem is less complicated, we consider it first. In this chapter we will review part of the paper GRÖTSCHEL, JÜNGER & REINELT (1991) where a preliminary version of the software was used for solving the problems.

11.1 Drilling of Printed Circuit Boards

One of the final steps in printed circuit board (PCB) production consists of the drilling of holes into the board. These holes are necessary for placing components (integrated circuits, transistors, etc.) onto the board or for realizing contacts between the different layers of the boards. A simplified view of such a board is given in Figure 11.1.

In general, holes are of different diameters. Hence drills have to be changed. In our application drills cannot be changed "on the fly". The head containing the drill has to move to an origin where the drill can be changed. Since time to change drills is considerable it only makes sense to drill all holes of one diameter before changing the drill.

The problem to be solved here is to find a sequence in which the holes are to be drilled such that the total production time is minimized. Following the remarks above, this problem decomposes into a set of single problems for each diameter.

We will now consider such a subproblem. Since the corresponding drill has to be loaded the machine has to move to the origin. Then all holes of the chosen diameter are drilled and the machine returns to the origin to start the next drilling sequence. The time to drill the holes cannot be influenced. Production time can only be decreased if positioning time, i.e., total time needed to reach the positions of the holes, is reduced. This amounts to solving a symmetric traveling salesman problem in the complete graph on $n+1$ nodes, if n holes are to be drilled (the node $n+1$ represents the origin).

The distance between two nodes i and j corresponds to the positioning time the machine needs to move from position i to position j, and therefore depends strongly on the machine characteristics. In practice, usually, this positioning time cannot be computed exactly. Positioning consists of three phases: accelerating the machine, running at full speed, slowing down to a complete stop. For small distances, full speed may not be reached and we may have anomalies in the sense that a farther position can be reached faster than a nearer position. Even if a timing function is available it may be not accurate or so complicated that its evaluation takes too long for large problem instances (where we cannot store a distance matrix). Therefore one has to be satisfied with making reasonable approximations on the true movement time.

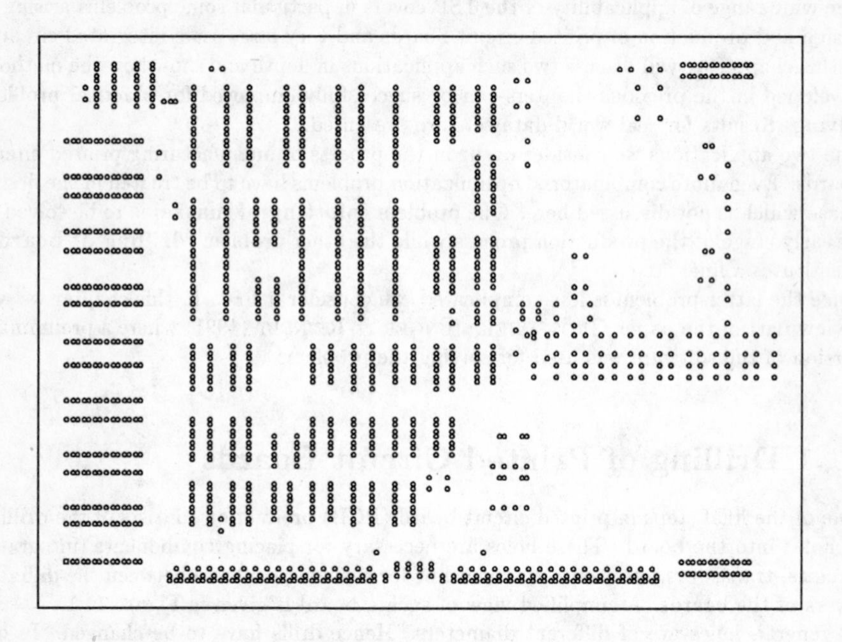

Figure 11.1 A printed circuit board

In the application treated in the following, the drilling head was moved by two motors running simultaneously in horizontal and vertical direction. Forgetting about machine characteristics, the longer of the two coordinate differences between two consecutive points in absolute value determines the positioning time. If we do not take into account acceleration and slowing down phase then the distance corresponds to the positioning time. In our example, the maximum distance between two points in the plane reflects the positioning time and can be taken as a reasonable approximation of the real situation. Note that, if the drilling head is moved by only one motor, then the Manhattan distance (if movement is only possible horizontally or vertically) or the Euclidean distance (if free movement is possible) can be appropriate.
The problem of minimizing drilling time for a printed circuit board therefore corresponds to solving a sequence of (unrelated) symmetric traveling salesman problem instances in

the plane where distances are given as maximum distances. We specify the distances in machine units. In the particular application, movement in horizontal direction was 10% faster than in vertical direction, but this can be easily taken into account by appropriate scaling.
Four problems of moderate sizes were examined to test our algorithms. Their characterizations are displayed in Table 11.2.

	da1	da2	da3	da4
Number of holes	2457	423	2203	2104
Number of drills	7	7	6	10
Size of problem 1	492	209	2114	656
Size of problem 2	1666	72	72	1302
Size of problem 3	156	51	2	1
Size of problem 4	122	66	3	117
Size of problem 5	16	14	4	3
Size of problem 6	4	5	8	7
Size of problem 7	1	6	–	7
Size of problem 8	–	–	–	4
Size of problem 9	–	–	–	6
Size of problem 10	–	–	–	1
Total length of moves in industry solution	3518728	1049956	1958161	4347902

Table 11.2 Characteristics of the drilling problems

These problems are not very large but there was one fact impeding the heuristics: For any printed circuit board, solutions had to be found within 5 minutes on a medium-sized computer.
This severe time restriction was met as follows. If the subproblem size was smaller than a limit of 300 nodes a nearest neighbor tour was computed. Otherwise the node reduction heuristic described in the preceding chapter was used to reduce the problem below this limit. For the representative TSP a farthest insertion tour was computed from which a tour for the original problem was derived (considering at most 200 insertion points).
To assess these solutions and to try to convince the engineers of spending more CPU time for optimizing the problems we also ran more elaborate heuristics, namely a simple Lin-Kernighan implementation (without node insertion and enumeration, and without use of efficient data structures) applied to a nearest neighbor tour.
We list the results obtained this way (CPU times are given for a SUN 3/60 workstation whose speed is about 3 MIPS).
Results published in GRÖTSCHEL, JÜNGER & REINELT (1991) are given in Tables 11.3 and 11.4. Relative comparison between two figures is always given as follows. Let a, b be two total lengths of positioning moves. Then the improvement yielded by b with respect to a is defined as $100 \cdot (a - b)/a$.

It should be remarked here that the industry solutions were already obtained by employing a heuristic.

	da1	da2	da3	da4
CPU time (min:sec)	1:58	0:05	1:43	1:43
Length of positioning moves	1695042	984636	1642027	1928371
Improvement w. r. t. to industry in %	51.83	6.22	16.14	55.14

Table 11.3 Previous results (fast version)

CPU time (min:sec)	96:03	6:27	169:15	70:03
Length of positioning moves	1517748	896669	1430556	1809556
Improvement w. r. t. to industry in %	56.87	14.60	26.94	58.38
Improvement w. r. t. to Table 10.3 in %	10.46	8.93	12.88	6.16

Table 11.4 Previous results

We repeated these experiments with the new implementations described in Chapters 6 and 7. Parameters were chosen based on the experience gained from previous computational experiments.
Since distances are given as maximum distances, the use of Voronoi diagrams for the L_∞-metric would be appropriate. However, we did not have a code for this Voronoi diagram at hand. Therefore we "approximated" computations using the L_2-metric diagram. Candidate subgraphs were computed taking nearest neighbors with respect to the Euclidean distance and also edges of the Delaunay graph for the L_2-metric Voronoi diagram.
We implemented a generic optimization routine which is called for every subproblem P. Depending on the size of the problem more or less work is spent for finding tours.

procedure optimize

(1) If the problem has less than three nodes compute the trivial solution and stop.

(2) Let n be the problem size and denote parameters as follows:
k number of nearest neighbors,
l number of moves that are examined in the Lin-Kernighan heuristic,
d maximal number of submoves for each move,
c number of alternate candidates for the first submove
Depending on the size of the problem set the parameters as follows.

(2.1) If $n < 100$ then set $k = 9$, $d = 15$, $l = 10 \cdot n$, $c = 2$.

(2.2) If $100 \leq n < 500$ then set $k = 8$, $d = 12$, $l = 10 \cdot n$, $c = 2$.

(2.3) If $500 \leq n < 1500$ then set $k = 7$, $d = 10$, $l = 10 \cdot n$, $c = 2$.

(2.4) If $n \geq 1500$ then set $k = 6$, $d = 10$, $l = 10 \cdot n$, $c = 1$.

(3) Compute the (Euclidean) k-nearest neighbor subgraph by enumeration if $n < 100$, otherwise by using the Delaunay graph.

(4) Use the nearest neighbor graph augmented by the edges of the Delaunay graph as candidate subgraph and switch to the maximum distance.

(5) Compute a nearest neighbor tour where the candidate edges are used to speed up computations.

(6) Perform the Lin-Kernighan heuristic (including node insertion moves) according to the parameter setting.

end of optimize

For assessing the quality of the tours found we also computed the simple 1-tree lower bound for every subproblem. Since the L_∞-metric Delaunay graph was not available we used Prim's algorithm to obtain a lower bound. The running time for the spanning tree computation is not included in Tables 11.5 and 11.6, because the computation takes time $O(n^2)$. If the L_∞-metric Voronoi diagram is available one can include this lower bound calculation into the problem solving code. L_∞-metric Delaunay graphs should be computable at least as fast as for the L_2-metric and their computation only slightly increases the overall running time.
In the first experiment we used a fast heuristic similar to the one used used to obtain the results of Figure 11.3.

procedure fastopt

(1) Partition the problem into subproblems.

(2) For every subproblem do the following.

(2.1) If the problem has fewer than 100 nodes then call *optimize*.

(2.2) If the problem has at least 100 nodes perform the node reduction heuristic with at least $m = \lceil \log \sqrt[4]{n} \rceil$ subdivisions and at most $\lceil \sqrt{n} \rceil / 2$ points in a bucket (see Chapter 8). For finding a global tour through the representative points call *optimize*. To reconstruct a tour for the original problem check $50 \cdot \log n$ insertion points.

end of fastopt

The results with this heuristic are documented in Table 11.5. We give comparisons with the tours provided by industry and with the tours computed by the fast heuristic of Table 11.3. To compare CPU times of the new experiments (obtained using a SUN SPARCstation SLC with 12.5 MIPS) with CPU times of the previous results we multiplied them by a factor of 4. For example, the running time for problem **da4** in Table 11.5 was about 26 seconds on a SPARCstation SLC, which in turn corresponds to about 6 seconds on a SPARCstation 10/20.

With about the same amount of CPU time results are considerably better than for the old version, in particular for problem da2. The quality guarantees derived from the 1-tree bound are about 20% which might be of interest for practitioners.

	da1	da2	da3	da4
CPU time (min:sec)	1:39	1:33	1:32	1:43
Length of positioning moves	1615031	844563	1537623	1878545
Improvement w. r. t. to industry in %	54.10	19.56	21.48	56.79
Improvement w. r. t. to Table 11.3 in %	4.72	14.23	6.36	2.58
Quality guarantee in %	21.24	21.22	20.55	19.66

Table 11.5 Results and comparison of fast heuristic

In the next experiment we applied the routine *optimize* to every subproblem. Results are displayed in Table 11.6. The tours obtained are compared with the industry solution, with the new fast heuristic (Table 11.5) and with the previously used heuristics (Tables 11.3 and 11.4).

	da1	da2	da3	da4
CPU time (min:sec)	3:15	1:32	1:59	3:33
Length of positioning moves	1440915	829484	1343011	1691869
Improvement w. r. t. to industry in %	59.05	21.00	31.41	61.09
Improvement w. r. t. to Table 11.3 in %	14.99	15.76	18.21	12.64
Improvement w. r. t. to Table 11.4 in %	5.06	7.49	6.12	6.50
Improvement w. r. t. to Table 11.5 in %	10.78	1.78	12.66	9.94
Quality guarantee in %	11.73	19.79	9.04	10.79

Table 11.6 Results and comparison of more elaborate heuristics

Table 11.6 exhibits a considerable improvement. The solutions found are still about 6% better than the solutions of Table 11.4. Quality guarantees are now in a range where one can safely conclude that the solutions computed are very good. But, what is even more surprising, the running time for the more sophisticated approach is only about double the CPU time of the fast heuristic. This shows that for problems of those moderate sizes we can still use sophisticated heuristics without spending too much CPU time.

Already with the solutions for da1 and da4 of Table 11.3, which are of rather moderate quality, production time could be reduced by 6%, respectively 10%. Since production time depends on the number of holes to be drilled, we cannot conclude from these figures to the possible reduction achievable if the heuristics documented in Table 11.6 were implemented.
But it is clearly seen that the TSP heuristics developed here can lead to substantial savings in industrial production processes. The heuristics are powerful enough to compute very good solutions in short time.
This final discussion leads to a side remark. Production time can also be reduced if the number of holes to be drilled is reduced. Many of these holes are just needed to connect wires on different layers, i.e., they are so-called vias. In the design phase for printed circuit boards (as well as for integrated circuits) one can apply combinatorial optimization to reduce the number of vias. For a report on this see GRÖTSCHEL, JÜNGER & REINELT (1989).

11.2 Plotting of PCB Production Masks

The second application we are concerned with deals with one of the first steps in the production process of printed circuit boards.
Complex printed circuit boards are usually produced by a photochemical process. For each layer of the board, the pattern of wires and contacts is produced by a sequence consisting of covering the board with light sensitive material, exposing this material to light, etching, cleaning, etc. The process is similar to the usual production of photographs. The structures that later on should appear on the board have been "drawn" on a mask (a negative) that is between the board and the light source so that certain parts of the board are not exposed to light. These unexposed areas will finally form the conductors, pads, and contacts of the layer. The question we address here is the generation of these masks.
The masks are made of glass and the patterns on the glass are generated optically using either ultraviolet light or laser beams. In our case, a photo plotter is used for the mask production.
The photo plotter works as follows. It has two modes, a "drawing mode" with which lines are plotted and a "flashing mode" to plot points. Points may be of various sizes and shapes and lines of different width. So, before plotting, an aperture has to be chosen that produces the required shape or width.
Points are plotted by moving the light source to certain coordinates on the board, choosing the aperture, and flashing the light. Lines are plotted by moving the head to one end of the line, choosing the aperture, opening the shutter, moving along the line with open shutter and closing the shutter at the end of the (not necessarily straight) line.
There is, given a pattern, nothing to be done about the time needed for drawing and flashing. This process requires a certain fixed time depending on the plotter characteristics. What can be optimized is the time needed for positioning head moves, i.e., moves of the head without drawing.

As for the drilling problem simultaneous movements in horizontal and vertical direction are possible to position the plotter head. This time, speeds were the same in both directions. Again, as no better means was available, maximum distances between points were taken to reflect positioning time. KORTE (1989) states an explicit function for computing the positioning time of a different plotter.
Figure 11.7 displays part of a mask from one of our application problems.

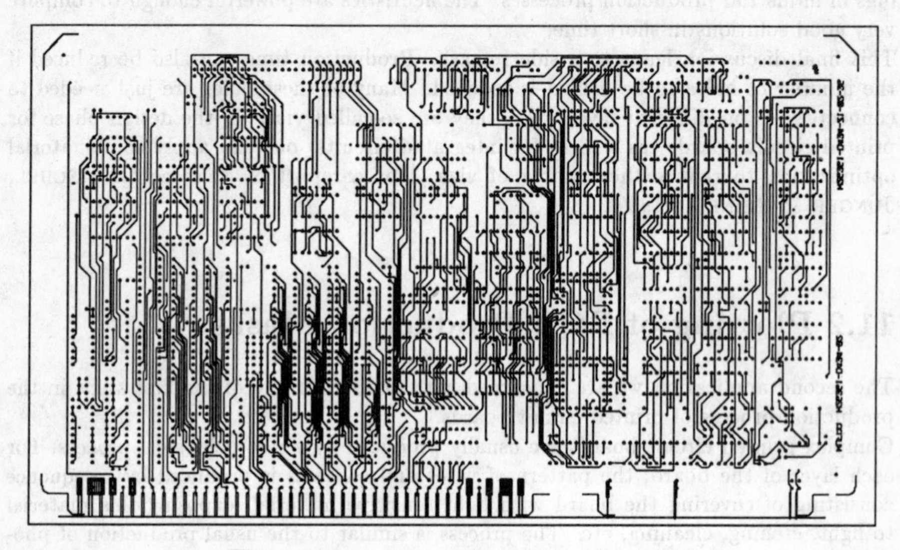

Figure 11.7 A mask for PCB production

In addition to the fact that not only points have to be flashed, but also lines have to be drawn, there are some further complicating points.

- There is no toolbox as in the drilling case. Apertures may be changed any time and also during movement.
- For small moves an aperture change may contribute to positioning time whereas it does not for long moves.
- Lines do not have to be drawn in one piece. In principle, preemptions are allowed, i.e., a line is plotted in several parts.

In particular, the last point makes the problem mathematically intractable. But it is also difficult to handle arbitrary aperture changes.
In our particular application we could discard at least some of these problems. Due to the required high accuracy of the plot, preemptions were not allowed and the number of aperture changes should be kept at a minimum.
Therefore we decided to decompose the plot into subplots where each subplot is composed of the structures to be plotted using the same aperture.
Still, there is the problem of determining the sequence in which the subplots are treated. Exact modeling of this problem seems to be hardly possible. Computation of the minimal length of positioning moves for a given aperture sequence involves the exact solution

of $\mathcal{NP}$-hard problems. Fortunately, in our problems only a few apertures were used. So it was clear in advance that the aperture sequence only had a negligible effect on the final result. In GRÖTSCHEL, JÜNGER & REINELT (1991) a nearest neighbor heuristic was applied to determine a sequence during the optimization process. In the experiments to be reported here we constructed the subproblems for every aperture and solved them in the order in which they were generated.
For every subproblem we have a starting point, namely the point where the previous subproblem solution terminated (for the first problem this is the origin). The final position reached in a subproblem solution is arbitrary. For the last problem we can include the origin and require that it is the final position to be reached.
Basically, we have therefore two types of subproblems, depending on whether an aperture for flashing or for drawing is selected.

Point flashing subproblem

Determine a shortest Hamiltonian path which starts at some given point and visits all points to be flashed. □

Line drawing subproblem

Starting at some given point determine a sequence in which the lines are plotted such that the total length of positioning moves is minimized. □

The point flashing problem is easily modeled. It amounts to solving a shortest Hamiltonian path problem with given starting node and unspecified (except for a final problem) terminating node. Modeling of path problems has been discussed already in Chapter 3. To avoid adding an artificial node we simplified the problem as follows. We computed a Hamiltonian tour and then removed the shorter one of the two edges between the starting node and its neighbors in the tour. The corresponding neighbor is then taken as endnode of the Hamiltonian path.
Treating the drawing subproblems is more complicated. The various aspects of modeling this problem which is a special case of the rural postman problem are discussed in GRÖTSCHEL, JÜNGER & REINELT (1991), an explicit transformation from the rural postman problem to the symmetric TSP is given in JÜNGER, REINELT & RINALDI (1994). Further transformations as well as computational results for a different problem set-up are presented in KORTE (1989).
We consider here a transformation particularly suited for the instances of the mask plotting problem we are faced with. A feature of the drawing subproblems occuring here is that only few lines touch. Most of the lines are isolated.
Suppose m lines are to be plotted. We also represent the starting point as a line (of length zero) and, if required, also the terminating point (the origin). It is straighforward to take a specified terminating point into account, so we assume in the following that the final position reached after having plotted all lines is arbitrary.
With each line i we associate two nodes i and $m+i$ (corresponding to the endpoints of the line) and represent the line drawing problem using the weighted complete graph $K_{2m} = (V, E)$ on $2m$ nodes. Nodes corresponding to the same coordinate are not identified. Edge weights c_{ij} are given as maximum distance between the points on the

printed circuit board represented by the nodes i and j. Edge weights $c_{i,m+i}$ are set to $-M$ (M a large positive number).
We then find a feasible solution for the line drawing subproblem by solving the traveling salesman problem in this graph. If M is large enough then all edges $(i, m+i)$ are contained in the solution and we can derive a sequence of machine moves for plotting the lines. If the edge $(1, m+1)$ represents the starting node, then either the neighbor of 1 (different from $m+1$) or the neighbor of $m+1$ (different from 1) will be selected as terminating point (depending on the length of the edge that can be eliminated). Note, that this does not necessarily lead to the best choice of the termination point. But, this slightly incorrect modeling will certainly have only negligible effect on the quality of the solutions of the practical problem.
Before describing our solution approach, we discuss the problem instances and list the results published using a preliminary implementation of the TSP software.
Five problem instances were available. Table 11.8 lists their respective characteristics and the properties of the industry solution which was obtained by a simple procedure based on sorting with respect to vertical coordinates.

	uni1	uni2	uni3	uni4	uni5
Number of drawn lines	6139	869	1360	49	38621
Length of lines	19123502	2102549	25552950	4761800	124351961
Number of flashes	2157	2496	1477	2478	1060
Number of apertures	7	9	5	5	5
Number of aperture changes	33	261	5	5	5
Length of moves in industry solution	41285752	26445205	77629210	38382300	296730563

Table 11.8 Characteristics of the plotting problems

Again, for this set of five problem instances, solutions were computed with a fast and with a more elaborate heuristic in GRÖTSCHEL, JÜNGER & REINELT (1991). CPU times are again from a SUN 3/60 workstation.
The first step for treating the five problems consists of a problem size reduction. In particular, problems uni1 and uni5 have a very large number of lines to be drawn. However, when looking closely at the respective plots we saw that many connections were realized as "wavy" lines. In order to achieve this form such lines were composed of many short (straight) lines connected in a zig-zag manner (this technique is used to obtain a denser packing of wires). It is reasonable to treat such lines as single lines and ignore that they really are composed of many very small segments. Having performed this reduction we now get the new plotting problems whose characteristics

are displayed in Table 11.9. (The sizes of the flashing subproblems have not changed, of course). Tables 11.10 and 11.11 give the results from GRÖTSCHEL, JÜNGER & REINELT (1991).

	uni1	uni2	uni3	uni4	uni5
Number of drawn lines	1411	41	355	18	610
Size of drawing problem 1	574	31	258	12	604
Size of drawing problem 2	63	10	87	2	2
Size of drawing problem 3	49	-	10	4	2
Size of drawing problem 4	725	-	-	-	-
Number of flashes	2157	2496	1477	2478	1060
Size of flashing problem 1	3	3	1432	2319	1060
Size of flashing problem 2	2152	136	45	159	-
Size of flashing problem 3	2	1817	-	-	-
Size of flashing problem 4	-	1	-	-	-
Size of flashing problem 5	-	1	-	-	-
Size of flashing problem 6	-	478	-	-	-
Size of flashing problem 7	-	60	-	-	-

Table 11.9 Characteristics of the reduced problems

In the new experiments, the flashing subproblems were treated in an analogous way as the drilling problems in the previous sections.

For the drawing subproblems we created a candidate subgraph in the same way as for the drilling problems (i.e., based on the Delaunay graph for the endpoints of the lines and nearest neighbor computations with respect to the Euclidean distance). In addition, we added the edges $(i, m+i)$, $1 \leq i \leq m$, to the candidate set. Recall that these edges have length $-M$.

We are now looking for a traveling salesman tour containing all edges $(i, m+i)$. Most of the construction heuristics cannot guarantee to find solutions satisfying this requirement. Here the nearest neighbor heuristic is appropriate. It will automatically generate

	uni1	uni2	uni3	uni4	uni5
CPU time (min:sec)	4:33	2:36	1:37	3:47	1:19
Number of aperture changes	7	9	5	5	5
Length of positioning moves	17731273	16345805	26870050	32935300	49737209
Improvement w. r. t. industry solution in %	57.0	38.19	65.39	14.19	83.24

Table 11.10 Previous results (fast version)

	uni1	uni2	uni3	uni4	uni5
CPU time (min:sec)	128:01	15:57	16:29	13:30	57:47
Number of aperture changes	7	9	5	5	5
Length of positioning moves	15611483	15337249	23499950	30871300	41270713
Improvement w. r. t. industry solution in %	62.19	42.00	69.73	19.57	86.09
Improvement w. r. t. Table 11.10 in %	11.96	6.17	12.54	6.27	17.02

Table 11.11 Previous results

a feasible tour, also if we speed it up using our candidate set because this set contains the critical edges. If we choose the number M carefully and large enough then none of our improvement heuristics will ever remove a critical edge from the current tour. In turned out that choosing M is indeed not always easy. In our application, for example, $M = 600,000$ was not big enough because some distances were in about this range.

In view of the results for the drilling problem we did not use bucketing any more in the new experiments. For the drawing subproblems it is not appropriate anyway. We only used the procedure *optimize* defined in the previous section on all subproblems.

In addition a lower bound was computed. For all subproblems a minimum weight spanning tree was computed. Since we had to use Prim's algorithm here, running times for lower bound computations are not included. Note that if an L_∞-metric Delaunay graph procedure is available then these computations only add a few percent to the total running time.

Results and comparisons with the previous experiments are given in Table 11.12. Again, the SUN SPARCstation SLC times were multiplied by 4 to obtain running times comparable with the SUN 3/60 times.

Using the new efficient algorithms the results documented in Tables 11.10 and 11.11 were outperformed by far. Even the solutions of Table 11.11 can be considerably improved in much less CPU time. By running the present code checking more moves in the Lin-Kernighan procedure or by running it several times on different starting solutions we believe that further reductions by 1–2% are possible (still in less time than in Table 11.11).

	uni1	uni2	uni3	uni4	uni5
CPU time (min:sec)	9:18	4:13	4:34	2:35	6:39
Number of aperture changes	7	9	5	5	5
Length of positioning moves	14386383	13998943	22555428	29766067	38056072
Improvement w. r. t. industry solution in %	65.15	47.06	70.94	22.45	87.17
Improvement w. r. t. Table 11.10 in %	18.86	14.36	16.0	9.62	23.49
Improvement w. r. t. Table 11.11 in %	7.85	8.73	4.02	3.58	7.79
Quality guarantee in %	16.49	9.70	13.31	2.87	22.71

Table 11.12 Results for the plotting problems

Of course, all tables only reflect the improvement in the length of the positioning moves measured in the maximum distance. The real effect on production time was only evaluated for the solutions given in Table 11.10. These solutions were compared with the industry solutions in real runs of the photo plotter. The decrease in the overall production time was 15.77%, 33.33%, 23.68%, 7.98%, and 8.78% for the respective problems. Still there is a considerable decrease in production time. The small decrease for problem uni5 of 8.78% compared to the enormous decrease in the length of the positioning moves is due to a large number of long wavy lines consisting of many small segments. This was also reflected by the reduction of the number of lines to be plotted from Table 11.8 to Table 11.9.
With the new solutions obtained here, production time can be reduced even further.

Chapter 12

Practical TSP Solving

In many practical applications, it is not required that true optimal solutions are computed. Possible reasons are for example:

- Due to an incorrect modeling of the real underlying problem, an optimal TSP solution may not correspond to an optimal solution for the application.
- Real time is not available to attempt to find an optimal solution.
- The size of the problem instances is too large for an exact optimization algorithm.

But, as theory improves and as ever more powerful hardware becomes available, the sizes of problem instances for which even optimal solutions can be found in reasonable time will also increase. Solving TSPs in practice should not mean to be satisfied with approximate solutions, but rather to try to find the best possible solution within the time that is available. If optimal solutions cannot be determined, then solutions should be delivered that are accompanied by a quality guarantee (which can be obtained by computing lower bounds). The fact, that the traveling salesman problem is $\mathcal{NP}$-hard, only means that for every algorithm there are difficult problem instances (provided $\mathcal{P}\neq\mathcal{NP}$), but it may well be the case that many problems arising in practice can be solved to optimality even if they are large.
It is beyond the purpose of this tract to discuss methods for the exact solution of the traveling salesman problem in depth. But, we want at least to indicate, how a hardware and software setup for the treatment of the TSP in practice should look like, that is able to find optimal solutions if time permits or that computes approximate solutions with certified quality guarantee. Detailed expositions for the TSP are JÜNGER, REINELT & THIENEL (1993) and JÜNGER, REINELT & RINALDI (1994), more general aspects are addressed in JÜNGER, REINELT & THIENEL (1994).

12.1 Determining Optimal Solutions

Every algorithm for finding exact solutions for larger TSP instances is built of methods for finding upper and lower bounds and of an enumeration scheme. For a given instance, lower and upper bounds (feasible solutions) are computed. In most cases, these bounds will not be equal, and therefore, only a quality guarantee for the feasible solution can be given, but optimality cannot be proved. An instance has then to be split into subproblems such that the union of feasible solutions of the subproblems gives the feasible solutions of the master problem. Subproblems are then processed in the same way. This approach can be visualized as generating a **branching tree** where each node corresponds to a problem and the sons of a node represent the subproblems into which it is split.

To make this approach practical, the most important task is to keep the generated tree small. This is only possible, if subproblems are solved at an early stage, i.e., at a low level of the branching tree, since we may have an exponential growth of the number of subproblems. A subproblem is solved if its upper and lower bound coincide or if its lower bound is above the best feasible tour found so far. Therefore, both determination of good tours and derivation of strong lower bounds are the keys to a successful search for optimal solutions.

In principle, an algorithm of this type can be composed of any lower bounding procedure and of any collection of heuristics. Whereas there has been a lot of work and progress in designing heuristics, the situation for lower bounds is not as satisfying.

We classify algorithms for finding the optimal solutions of the traveling salesman problem as follows.

Branch & bound algorithms

Lower bounds are derived by purely combinatorial means, i.e., discrete relaxations of the TSP are solved using discrete methods.

□

Branch & cut algorithms

Lower bounds are obtained from linear programming relaxations.

□

An abundant number of branch & bound algorithms that are based on the relaxations mentioned in Chapter 10 have been designed and implemented. Most approaches use the 1-tree or the 2-matching relaxation for the symmetric TSP and the assignment relaxation for the asymmetric TSP (BALAS & TOTH (1985), MILLER & PEKNY (1991), MILLER, PEKNY & THOMPSON (1991)).

Yet, up to date all these algorithms are outperformed by branch & cut approaches. Prime contributions in this area, which laid the foundations for successful TSP solving, are GRÖTSCHEL (1977), CROWDER & PADBERG (1980), PADBERG & RINALDI (1991) and GRÖTSCHEL & HOLLAND (1991) which also visualize the enormous progress that has been made in the recent past. Since the same upper bounding heuristics are used in both branch & bound and branch & cut methods, the success of the latter approach depends on the superiority of the linear programming bounds.

Linear programming lower bounds are obtained by optimizing the objective function over a polytope P such that $P_T \subseteq P$ (where P_T is the traveling salesman polytope). The strongest such relaxation we have discussed so far is the subtour relaxation. It turns out that this relaxation is not sufficient for successful problem solving. This is also the reason why branch & bound algorithms based on 1-trees are quite limited in their ability to solve problems to optimality.

What is needed is a deeper knowledge about the polytope P_T. More exactly, we are interested in knowledge about the facet structure of P_T. Since the traveling salesman problem is $\mathcal{NP}$-hard, there is no hope of finding the complete description of P_T with linear equations and inequalities. But, even though the complete characterization cannot be determined, it is reasonable to look for classes of facets because every new class leads to a stronger relaxation.

To date, many classes of facets of P_T are known (see GRÖTSCHEL & PADBERG (1985) and NADDEF & RINALDI (1991,1993) for extensive discussions), among them are **subtour elimination constraints, 2-matching constraints, comb constraints**, or **clique tree constraints**. The degree equations define the affine hull of P_T, so also its dimension $\binom{n}{2} - n$ is known.

To get an impression of the exploding complexity of the facet structure of P_T, we mention that the traveling salesman polytope on 7 nodes has 3,437 facets (BOYD & CUNNINGHAM (1991)). Using a computer code, CHRISTOF, JÜNGER & REINELT (1991) show that for 8 nodes the number of facets is already 194,187. These figures exhibit a further difficulty that arises when trying to solve LP relaxations using linear programming techniques: the number of inequalities is simply too large to be listed explicitly.

This difficulty is overcome by generating inequalities only "as needed" using the so-called **cutting plane approach**. We give the principle idea.

procedure cutting_plane

(1) Select an initial polytope $P \supseteq P_T$, e.g., $\{x \in \mathbf{Q}^{\binom{n}{2}} \mid 0 \leq x_{ij} \leq 1, \text{ for all } ij \in E_n\}$ or the fractional 2-matching polytope.

(2) Repeatedly perform the following steps.

(2.1) Solve the linear programming problem $\min\{c^T x \mid x \in P\}$ to obtain an optimal solution x^*.

(2.2) If x^* is the incidence vector of a tour, then stop (an optimal tour is found).

(2.3) Find an inequality $a^T x \leq \alpha$ which is valid for P_T, but not satisfied by x^*, i.e., $a^T x^* > \alpha$.

(2.4) If no such inequality could be found, then stop (the best possible bound is determined).

(2.5) Add the inequality found in Step (2.3) to the current linear program.

end of cutting_plane

The hope is to terminate this procedure in Step (2.2), since then the traveling salesman problem is solved. But, because we do not know all facets describing P_T and because we are not able to identify all violated ones from the classes that are known, we will very likely terminate in Step (2.4). In that case, we have to follow the branch & bound approach and split the current problem under examination into subproblems according to some scheme. For solving the subproblems, the same cutting plane procedure is used. Note that all inequalities derived for some problem remain valid for the subproblems. Since this approach is different from the usual branch & bound, Padberg and Rinaldi coined the name "**branch & cut**" for this procedure.

In the core of the cutting plane approach there is the problem of finding violated inequalities in Step (2.3). This is not only intuitively clear, but also verified from theory as we will indicate now.
Let $Ax \leq b$ be a system of inequalities and consider the following problem.

Separation problem

Given a rational vector y decide if $Ay \leq b$ or not. If this is not the case then provide an inequality of this system that is violated by y.

□

Note, that this is not a trivial problem, because we may not have an explicit list of the inequalities or it may require exponential time to list all inequalities. It is a fundamental theorem (GRÖTSCHEL, LOVÁSZ & SCHRIJVER (1988)) that the linear programming problem $\min\{c^T x \mid Ax \leq b\}$ is solvable in polynomial time if and only if the separation problem for $Ax \leq b$ is solvable in polynomial time (with appropriate complexity measures).
Therefore, a basic ingredient to make the cutting plane approach work for solving traveling salesman problems is to have powerful procedures for generating violated inequalities in Step (2.3). Though many facet defining inequalities are known for P_T, only for a small minority there are efficient separation heuristics or exact algorithms.
It is surprising enough that indeed large problems with several thousand nodes can be solved to optimality by branch & cut.
Concerning the implementation of a branch & cut procedure, several issues have to be addressed. In the following we give a list of some aspects that have to be taken into consideration.

- A first observation is that it is not possible to keep all variables (edges) explicitly in the linear program. For example, for problem pr2392 there are 2,859,636 variables to be considered. To overcome this problem, one only keeps a set of active variables and then uses reduced cost pricing to price out the inactive variables. If they do not price out correctly, they have to be added to the active set and the corresponding columns have to be generated.
- Usually, there are many heuristics or exact algorithms for separation available. It has to be decided which routines are executed how often. To speed up computations one should use heuristics as long as possible and then switch to exact separation.
- In the first phases of the cutting plane procedure many inequalities might be generated. One has to decide if part or all of them are added to the current relaxation. Criteria according to which inequalities are selected have to be specified.
- One has to choose an LP package for the efficient treatment of the linear programs. So far, only implementations of the Simplex algorithm proved to be appropriate here, since they allow easy addition of constraints and computation of reduced costs.
- If it comes to branching, a branching strategy has to be developed. Normally, one generates two subproblems according to fixing a selected variable to 0 or to 1. A rule for selecting this variable has to be found.

- To keep the LP size handy, one occasionally must reduce it by eliminating constraints. A reasonable criterion is to eliminate those constraints that are not binding at the current optimal solution. However, during the optimization process eliminated inequalities may become necessary again. So it can be useful to keep eliminated inequalities in storage and check their violation after each phase (in particular, if it took a long time to identify them).
- To prevent the search tree from growing too large, very good upper bounds, i.e., feasible solutions, have to be computed. Moreover, if the gap between upper and global lower bound for a problem becomes small enough, then it is possible to eliminate variables permanently based on reduced costs. Lower bounds obtained at some non-root node are only valid for the subproblem represented by that node. Such bounds can be used for fixing variables temporarily.
- When solving a relaxation problem using cutting planes, bounds usually improve fast in the first cutting plane phases. After some time, however, the increase becomes very small. One has to decide if it is preferrable to suspend working on the current subproblem or to enforce a branching step.
- In the branching tree, several nodes are active in the sense that their corresponding subproblem is not yet solved. The node (or nodes) which are to be worked on next have to be selected.
- Suitable data structures have to be developed for storing inequalities in a compact way. In particular, one has to be able to check whether an active variable occurs in some inequality or not. This is important for performing an efficient reduced cost pricing.

There are no best solutions to the questions listed here. Answers are always problem dependent and can only be based on experience with the branch & cut procedure for a specific problem.
We think that some further quality can be gained if we take interactions between lower and upper bounding procedures into account. In the next section we will present an implementation concept that supports such interactions.

12.2 An Implementation Concept

We now address the question of how an implementation of traveling salesman algorithms should look like. We will present a concept for a hardware and software setup to solve traveling salesman problems in practice.
What should a practical TSP package accomplish? To our opinion, it should be designed according to the principle: "Find the best possible traveling salesman tour within the available CPU time and within the limits given by the hardware and supply it with a quality guarantee". This principle covers three particular situations.

(i) If CPU time is a scarce resource, then it must be possible to obtain reasonable solutions even for large problem instances.

(ii) If moderate CPU time is available then, depending on the problem size, (proved) qualities of 2–10% should be achieved. The real quality should be between 1% and 5%.

(iii) If CPU time is virtually unlimited, then the package must have a possible path to optimality, i.e., must be capable of finding an optimal solution (in principle).

In this view, an optimal solution for a problem instance is not the shortest tour, but rather the best tour that can be computed obeying the limits imposed by the production environment. The shortest tour, if it is found after some deadline was missed, is of no value for the application.

The hardware setup we are going to describe is motivated by the fact that today's computer equipment does not limit the software developer any more. Even powerful hardware is cheap, huge main memory sizes and virtual storage management allow almost arbitrary program sizes. Also disk space is available as needed. A new feature that can (and has to) be employed is parallelism, be it at a workstation level (where we have clusters of workstations) or at a processor level (where a parallel processor consists of several thousands of microprocessors). New architectures are available that will influence algorithm design.

Discussion in previous chapters has shown that we need three basic ingredients to meet the goals stated at the beginning of this chapter:

- an **upper bound component** (comprising heuristics of all kinds for finding approximate solutions quickly),
- a **lower bound component** (necessary for proving optimality or deriving quality guarantees),
- and a **branch & cut component** (providing the possibility to find optimal solutions).

Since the user of this software cannot be expected to have the knowledge to exploit the potential of the software, we have provided in our setup a supervisor component that guides the optimization process and coordinates interaction between the various algorithms.

Figure 12.1 gives a schematic view of the components of the system showing the three basic components.

The supervisor component is responsible for guiding the optimization and controlling the flow of information between the various components. It analyzes the problem instance to be treated and, based on the given time constraints, it selects algorithms to be performed. The supervisor constitutes kind of an expert system that has knowledge about the CPU time requirement and the expected quality of the various heuristics depending on characteristics of the problem instance (Euclidean, metric, geographical, clustered configuration, number of nodes, etc.).

In this text we have described a lot of experiments that can be used to initialize the data base of the supervisor. In the ideal case, the supervisor has a learning ability that adapts the internal data base according to acquired knowledge when solving problems. In particular, the supervisor decides whether the branch & cut part is activated at all.

The heuristic component selects heuristics as needed in the time range given by the supervisor. It must contain all types of heuristics we discussed so far to be prepared to find reasonable tours fast or very good tours in moderate time. It has to be able to treat problems of virtually any size. The heuristic component may also contain special auxiliary parts that are used by several heuristics. This is indicated here by specifying components for computing the convex hull or the Delaunay triangulation.

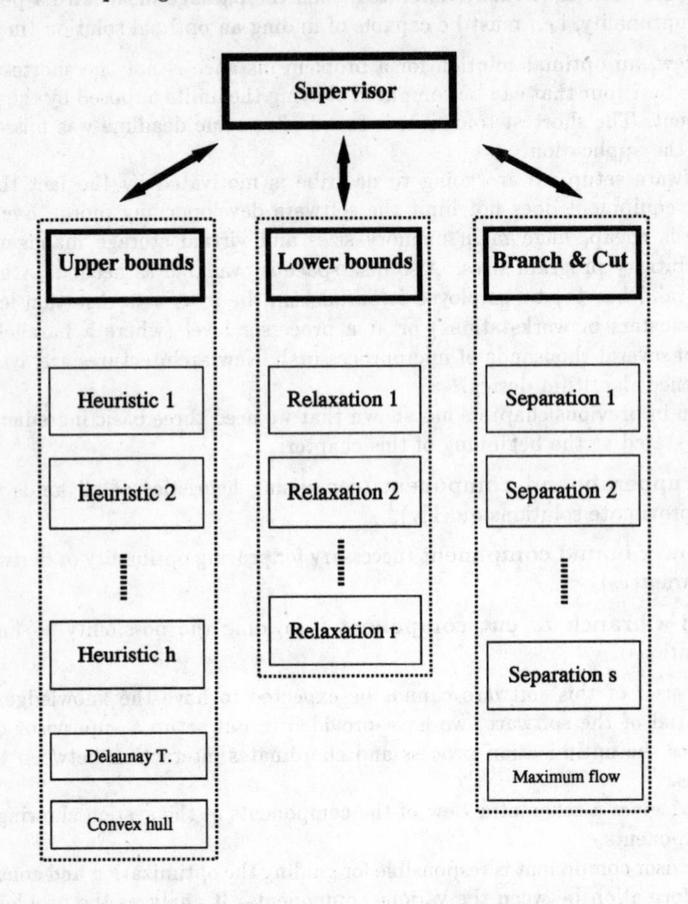

Figure 12.1 Components of the algorithmic framework

Besides the branch & cut part, we also have another component for providing lower bounds. As discussed in Chaper 10, there are many possibilities for deriving lower bounds of different qualities. This is of importance if the branch & cut component cannot be used due to a too large problem size. If branch & cut can be applied then it usually makes no sense to use this component since the linear programming bounds will be superior.

The third basic component constitutes the branch & cut part which should be activated whenever possible. It has to contain a powerful LP solver because a major part of the

CPU time will be spent for solving linear programs. Furthermore, it should provide as many separation algorithms as are available, comprising fast heuristics for simple inequalities, more elaborate heuristics for complicated inequalities, and exact separation routines. Though much is known about exact separation, there still is need for further research. Separation is the crucial part for extending capabilities of present codes. Also in this part, we may have auxiliary routines needed by many separation procedures. One example is maximum flow computation which is heavily used in TSP separation algorithms and heuristics.

We have intentionally drawn the various components and subcomponents separately. The reason is that we have a high degree of parallelism in this concept. All heuristics can be performed in parallel. The same is true for relaxation procedures or for the separation components. In particular, it should be annotated that separation procedures do not have to provide inequalities derived for the current LP. Since LPs do not change very much in later phases, an inequality derived for some previous LP might as well be violated in the current LP. Therefore we may have an identical program running several times in parallel, each supplied with a different fractional solution.

For the technical realization of this concept, we have a cluster of workstations or a transputer network in mind where each box of Figure 12.1 is assigned to one processor. Based on the available hardware, the supervisor can decide how to exploit it. If many processors are at hand, boxes may be duplicated several times. We can also make use of special parallel architectures to implement special purpose algorithms. So the concept permits employment of different hardware in an integrated framework.

12.3 Interdependence of Algorithms

In this final section we will indicate that the three components of our concept are not independent, but, on the contrary, exchange of information adds to problem solving capabilities. We will discuss some examples showing that communication between the various algorithmic components is highly desirable.

12.3.1 Selecting an Initial Set of Active Variables

As noted above, we cannot perform branch & cut for large problems keeping the full set of variables in the linear programs. A reasonable initial set of active variables is, for example, given by the k nearest neighbor graph. Using the Delaunay graph we can compute this set very efficiently for Euclidean problems. If the nodes form several clusters then the nearest neighbor subgraph is not connected. In such a case we can augment it by adding the edges of the Delaunay graph. To make sure that the active set contains a Hamiltonian tour one can add the edges of a tour found by some heuristic. Another possibility is to use several heuristics to compute a set of tours and then take the union of all tour edges as the initial set.

□

12.3.2 Augmenting the Set of Active Variables

During the algorithm, we will reach the point where all active variables price out correctly. Then the reduced costs of all variables have to be checked and some (or all) of them have to be introduced into the current linear program. Since this full pricing is very time consuming, it is reasonable to have heuristics run in parallel that compute very good tours and, hence, can provide the cutting plane part with promising new active variables.

□

12.3.3 Augmenting the Candidate Subgraphs for Heuristics

Also for the heuristics, we have the problem of determining a reasonable candidate subgraph. If reduced cost pricing introduces a variable into the linear program then it should also appear in the candidate set. Therefore, both components can share information about promising edges.

□

12.3.4 Driving the Heuristics

Possibly never a linear program solved during the branch & cut algorithm will yield an incidence vector of a tour as its optimal solution. Nevertheless, it can provide information on how parts of an optimal tour could look like. Figure 12.2 shows the LP solution of the subtour relaxation for problem **dantzig42** (solid edges correspond to variables with value 1, broken edges to variables with value $\frac{1}{2}$).

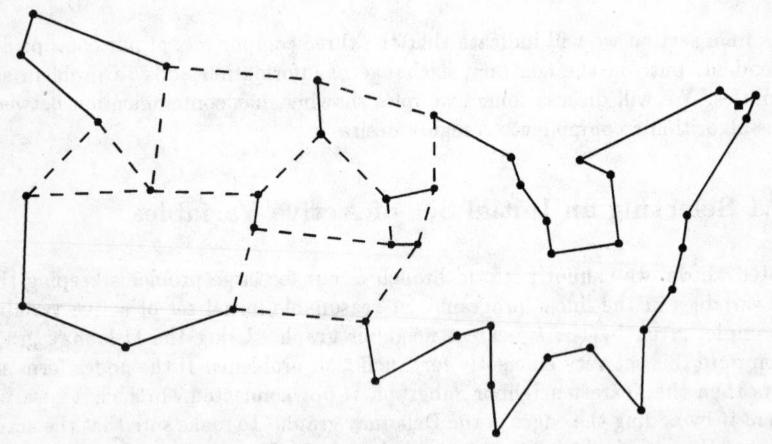

Figure 12.2 A fractional solution for problem **dantzig42**

The complete right part of this solution is identical to the optimal tour. We therefore exploit fractional LP solutions to drive heuristics. This is done in the following way.

procedure exploit_lp

(1) Sort the variables appearing with positive value in the current LP solution according to their value (if they have equal values they are sorted with respect to increasing objective function coefficient).

(2) Scan these edges and select an edge if it does not create a cycle with the edges selected so far. This is essentially the greedy algorithm and will yield a system of paths.

(3) Apply the Clarke-Wright savings heuristic to connect these paths in a cheap way.

(4) Improve the current tour by employing the Lin-Kernighan heuristic.

(5) Add the tour edges to the current candidate subgraph and add also the corresponding variables to the current linear program (if they were not yet taken into account).

end of exploit_lp

□

Finally, we show results of two experiments visualizing the above connections. To this end we have employed the branch & cut implementation of JÜNGER, REINELT & THIENEL (1993) to solve the relaxation of the TSP consisting of subtour elimination and 2-matching constraints. After the solution of each LP we have performed the heuristic *exploit_lp* to construct tours based on the current LP solution. Exchange of knowledge about reasonable edges was performed as in 12.3.2 and 12.3.3. The initial active set consisted of the six nearest neighbors.

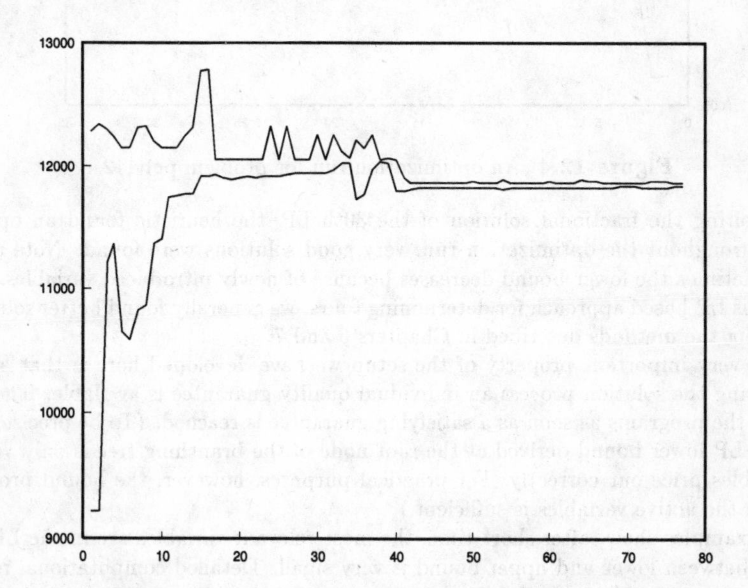

Figure 12.3 An optimization run for problem fl417

Figure 12.3 shows the development of upper and lower bounds obtained this way for problem fl417. The horizontal axis gives the number of the LP that is solved and the vertical axis gives lower and upper bounds. The optimal tour length is 11861. Note in particular the LPs where the lower bound drops considerably. At such points, the heuristic introduced new edges into the LP.

Problem fl417 is difficult because it has several clusters of points and it is therefore not clear which connections between the clusters should be chosen. Here the heuristic provides the necessary edges.

The second example is a run for problem pcb442 (Figure 12.4). Here the optimal solution value 50778 is depicted as horizontal line in the figure.

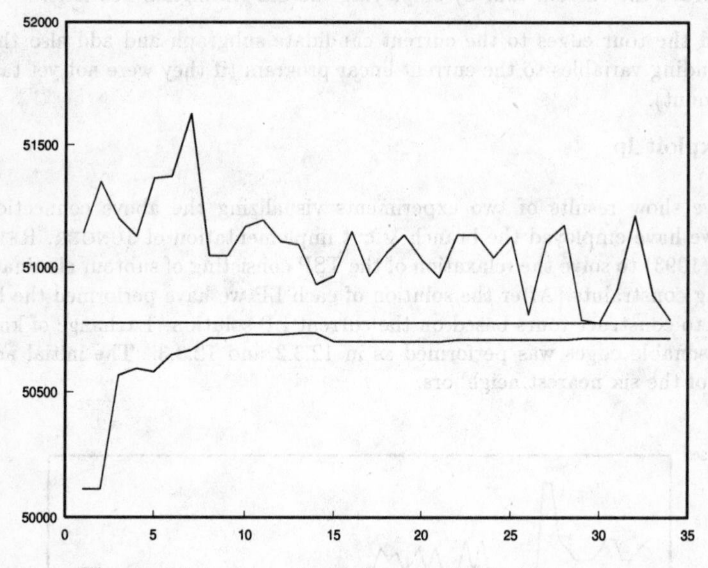

Figure 12.4 An optimization run for problem pcb442

By exploiting the fractional solution of the 30th LP, the heuristic found an optimal tour. Throughout the optimization run, very good solutions were found. Note again, that sometimes the lower bound decreases because of newly introduced variables.

Using this LP based approach for determining tours, we generally found better solutions than using the methods described in Chapters 6 and 7.

Another very important property of the setup we have developed here is that at any time during the solution process an individual quality guarantee is available. The user can halt the programs as soon as a satisfying guarantee is reached. (To be precise, note that the LP lower bound derived at the root node of the branching tree is only valid if all variables price out correctly. For practical purposes, however, the bound provided based on the active variables is sufficient.)

As the examples show, after short time, the most relevant variables are in the LP and the gap between lower and upper bound is very small. Detailed computational results are documented in JÜNGER, REINELT & THIENEL (1993).

Appendix

TSPLIB

All our sample problem instances have been taken from the library TSPLIB, which is a publicly available set of TSP and vehicle routing problem data. It comprises most of the problem instances for which computational results have been published. The reader is invited to conduct own experiments with this test set. A detailed description is REINELT (1991a).

Access

TSPLIB is electronically distributed and is available at Konrad-Zuse-Zentrum für Informationstechnik Berlin (ZIB) where data for various mathematical programming problems is collected. We give the instructions for using the `eLib` service at ZIB.
Data files are arranged in a two level directory structure. There is an `index` file in each directory that contains names of available files or subdirectories, `info` files contain descriptions of data formats, `readme` files containing more information are usually included in specific problem libraries.
All files can be obtained by e-mail, by anonymous ftp, or interactively via the ZIB electronic mail library service `eLib`. The respective addresses are:

E-mail:

elib@ZIB-Berlin.de

Datex-P:

+45050939033 (WIN)
+2043623939033 (IXI)

Internet:

telnet elib.zib-berlin.de (130.73.108.11)
rlogin elib.zib-berlin.de (130.73.108.11) (login as elib; no password is required)
gopher elib.zib-berlin.de (130.73.108.11)
anonymous ftp elib.zib-berlin.de (130.73.108.11)

In remote dialogue mode, `eLib` provides a command line interface and selection menus for browsing files. File selections are offered in top-down fashion according to directory structure. Select `MP-TESTDATA` at the top level menu displayed after login. Use the SEND command for obtaining a local copy of a file, it will be sent to you by e-mail in response. When using ftp, cd to `pub/mp-testdata/tsp`.
Help information on how to use `eLib` is provided online or can be obtained by sending a mail just containing "help" to `elib@ZIB-Berlin.de`.

Status of Problems

We give the current status of the symmetric TSP instances in TSPLIB. Distance types are described in REINELT (1991a)

Name	#cities	Type	Bounds
ali535	535	GEO	**202310**
att48	48	ATT	**10628**
att532	532	ATT	**27686**
bayg29	29	GEO	**1610**
bays29	29	GEO	**2020**
bier127	127	EUC_2D	**118282**
brazil58	58	MATRIX	**25395**
brd14051	14051	EUC_2D	[465044,479357]
burma14	14	GEO	**3323**
d198	198	EUC_2D	**15780**
d493	493	EUC_2D	**35002**
d657	657	EUC_2D	**48912**
d1291	1291	EUC_2D	**50801**
d1655	1655	EUC_2D	**62128**
d2103	2103	EUC_2D	[79743,80259]
d18512	18512	EUC_2D	[644470,651320]
dantzig42	42	MATRIX	**699**
dsj1000	1000	EUC_2D	**18659688**
eil51	51	EUC_2D	**426**
eil76	76	EUC_2D	**538**
eil101	101	EUC_2D	**629**
fl417	417	EUC_2D	**11861**
fl1400	1400	EUC_2D	[19849,20127]
fl1577	1577	EUC_2D	[22137,22249]
fl3795	3795	EUC_2D	[28594,28772]
fnl4461	4461	EUC_2D	**182566**
gil262	262	EUC_2D	**2378**
gr17	17	MATRIX	**2085**
gr21	21	MATRIX	**2707**
gr24	24	MATRIX	**1272**
gr48	48	MATRIX	**5046**
gr96	96	GEO	**55209**
gr120	120	MATRIX	**6942**
gr137	137	GEO	**69853**
gr202	202	GEO	**40160**
gr229	229	GEO	**134602**
gr431	431	GEO	**171414**
gr666	666	GEO	**294358**
hk48	48	MATRIX	**11461**
kroA100	100	EUC_2D	**21282**
kroB100	100	EUC_2D	**22141**
kroC100	100	EUC_2D	**20749**
kroD100	100	EUC_2D	**21294**
kroE100	100	EUC_2D	**22068**

Table 13.1 Symmetric traveling salesman problems (Part I)

Name	#cities	Type	Bounds
kroA150	150	EUC_2D	**26524**
kroB150	150	EUC_2D	**26130**
kroA200	200	EUC_2D	**29368**
kroB200	200	EUC_2D	**29437**
lin105	105	EUC_2D	**14379**
lin318	318	EUC_2D	**42029**
linhp318	318	EUC_2D	**41345**
nrw1379	1379	EUC_2D	**56638**
p654	654	EUC_2D	**34643**
pcb442	442	EUC_2D	**50778**
pcb1173	1173	EUC_2D	**56892**
pcb3038	3038	EUC_2D	**137694**
pla7397	7397	CEIL_2D	[23076619,23262472]
pla33810	33810	CEIL_2D	[65667327,66138592]
pla85900	85900	CEIL_2D	[141603586,142514146]
pr76	76	EUC_2D	**108159**
pr107	107	EUC_2D	**44303**
pr124	124	EUC_2D	**59030**
pr136	136	EUC_2D	**96772**
pr144	144	EUC_2D	**58537**
pr152	152	EUC_2D	**73682**
pr226	226	EUC_2D	**80369**
pr264	264	EUC_2D	**49135**
pr299	299	EUC_2D	**48191**
pr439	439	EUC_2D	**107217**
pr1002	1002	EUC_2D	**259045**
pr2392	2392	EUC_2D	**378032**
rat99	99	EUC_2D	**1211**
rat195	195	EUC_2D	**2323**
rat575	575	EUC_2D	**6773**
rat783	783	EUC_2D	**8806**
rd100	100	EUC_2D	**7910**
rd400	400	EUC_2D	**15281**
rl1304	1304	EUC_2D	**252948**
rl1323	1323	EUC_2D	**270199**
rl1889	1889	EUC_2D	**316536**
rl5915	5915	EUC_2D	[563416,565585]
rl5934	5934	EUC_2D	[554070,556146]
rl11849	11849	EUC_2D	[920847,923473]
st70	70	EUC_2D	**675**
swiss42	42	MATRIX	**1273**
ts225	225	EUC_2D	**126643**
u159	159	EUC_2D	**42080**
u574	574	EUC_2D	**36905**
u724	724	EUC_2D	**41910**
u1060	1060	EUC_2D	**224094**
u1432	1432	EUC_2D	**152970**
u1817	1817	EUC_2D	**57201**
u2152	2152	EUC_2D	[64163,64294]
u2319	2319	EUC_2D	[234256,234519]
vm1084	1084	EUC_2D	**239297**
vm1748	1748	EUC_2D	**336556**

Table 13.1 Symmetric traveling salesman problems (Part II)

References

The citations of each publication are given between brackets.

E.H.L. Aarts & J. Korst (1989a), *Simulated Annealing and Boltzmann Machines: A Stochastic Approach to Combinatorial Optimization and Neural Computing*, John Wiley & Sons, Chichester. [155]

E.H.L. Aarts & J. Korst (1989b), "Boltzmann Machines for Traveling Salesman Problems", *European Journal of Operations Research 39,* 79–95. [155]

P. Ablay (1987), "Optimieren mit Evolutionsstrategien", *Spektrum der Wissenschaft 7,* 104–115. [157]

G. Akl & G.T. Toussaint (1978), "A Fast Convex Hull Algorithm", *Information Processing Letters 7,* 219–222. [58, 62]

I. Althöfer & K.-U. Koschnick (1989), "On the Convergence of Threshold Accepting", Research Report, Universität Bielefeld. [156]

D. Applegate, R.E. Bixby, V. Chvátal & W. Cook (1991,1993), Personal Communication [1]

D. Applegate, V. Chvátal & W. Cook (1990), "Data Structures for the Lin-Kernighan Heuristic", Talk presented at the TSP-Workshop 1990, CRPC, Rice University. [23]

J.L. Arthur & J.O. Frendeway (1985), "A Computational Study of Tour Construction Procedures for the Traveling Salesman Problem", Research Report, Oregon State University, Corvallis. [73]

E. Balas (1989), "The Prize Collecting Traveling Salesman Problem", *Networks 19,* 621–636. [35]

E. Balas & P. Toth (1985), "Branch and Bound Methods", in: E.L. Lawler, J.K. Lenstra, A.H.G. Rinnooy Kan & D.B. Shmoys (eds.) *The Traveling Salesman Problem*, John Wiley & Sons, Chichester, 361–401. [176, 186, 201]

M.L. Balinski (1970), "On Recent Developments in Integer Programming", in: H.W. Kuhn (ed.) *Proceedings of the Princeton Symposium on Mathematical Programming*, Princeton Univ. Press, Princeton, 267–302. [180]

J.J. Bartholdi & L.K. Platzman (1982), "An $O(n \log n)$ Planar Travelling Salesman Heuristic Based on Spacefilling Curves", *Operations Research Letters 4,* , 121–125. [133, 135]

L.J. Bass & S.R. Schubert (1967), "On Finding the Disc of Minimum Radius Containing a Given Set of Points", *Math. Computation 21,* 712–714. [55]

J. Beardwood, J.H. Halton & J.M. Hammersley (1959), "The Shortest Path Through Many Points", *Proc. Cambridge Philos. Society 55,* 299–327. [135]

M. Bellmore & S. Hong (1974), "Transformation of Multisalesmen Problem to the Standard Traveling Salesman Problem", *Journal of the ACM 21,* 400–504. [33]

J.L. Bentley (1990), "K-d-trees for Semidynamic Point Sets", Sixth Annual ACM Symposium on Computational Geometry, Berkeley, 187–197. [67]

J.L. Bentley (1992), "Fast Algorithms for Geometric Traveling Salesman Problems", *ORSA Journal on Computing 4,* 387–411. [73, 83, 106, 112, 114, 119]

J.L. Bentley & M.I. Shamos (1978), "Divide and Conquer for Linear Expected Time", *Information Processing Letters 7,* 87–91. [57, 62]

R.E. Bixby (1994), "Progress in Linear Programming", *ORSA Journal on Computing 6,* 15–22. [13]

R.E. Bland & D.F. Shallcross (1989), "Large Traveling Salesman Problems Arising from Experiments in X-ray Crystallography: A Preliminary Report on Computation", *Operations Research Letters 8*, 125–128. [36]

K.H. Borgwardt, N. Gaffke, M. Jünger & G. Reinelt (1991), "Computing the Convex Hull in the Euclidean Plane in Linear Expected Time", *The Victor Klee Festschrift*, DIMACS Series in Discrete Mathematics and Theoretical Computer Science 4, 91–107. [60, 62]

S.C. Boyd & W.H. Cunningham (1991), "Small Traveling Salesman Polytopes", *Mathematics of Operations Research 16*, 259–271. [202]

S.C. Boyd & W.R. Pulleyblank (1990), "Optimizing over the Subtour Polytope of the Traveling Salesman Problem", *Mathematical Programming 49*, 163–188. [163, 170]

R.E. Burkard (1990), "Special Cases of Travelling Salesman Problems and Heuristics", *Acta Math. Appl. Sinica 6*, 273–288. [11]

G. Carpaneto & P. Toth (1983), "Algorithm for the Solution of the Assignment Problem for Sparse Matrices", *Computing 31*, 83–94. [167]

G. Carpaneto, M. Fischetti & P. Toth (1989), "New Lower Bounds for the Symmetric Travelling Salesman Problem", *Mathematical Programming 45*, 233–254. [186]

V. Cerny (1985), "A Thermodynamical Approach to the Travelling Salesman Problem: An Efficient Simulation Algorithm", *Journal on Optimization Theory and Applications 45*, 41–51. [155]

T. Christof, M. Jünger & G. Reinelt (1991), "A Complete Description of the Traveling Salesman Polytope on 8 Nodes", *Operations Research Letters 10*, 497–500. [202]

N. Christofides (1976), "Worst Case Analysis of a New Heuristic for the Travelling Salesman Problem", Research Report, Carnegie-Mellon University, Pittsburgh. [91]

B.A. Cipra (1993), "Quick Trips", SIAM News, Oct. 1993. [1]

G. Clarke & J.W. Wright (1964), "Scheduling of Vehicles from a Central Depot to a Number of Delivery Points", *Operations Research 12*, 568–581. [94]

N.E. Collins, R.W. Eglese & B.L. Golden (1988), "Simulated Annealing: An Annotated Bibliography", *American Journal Math. and Management Science 8*, 205–307. [155]

S.A. Cook (1971), "The Complexity of Theorem-Proving Procedures", Proc. 3rd Annual ACM Symp. Theory of Computing, 151–158. [7]

T.H. Cormen, Ch.E. Leiserson & R.L. Rivest (1989), *Introduction to Algorithms*, MIT Press, Cambridge. [14, 18, 25, 27]

G. Cornuejols, J. Fonlupt & D. Naddef (1985), "The Traveling Salesman Problem on a Graph and some Related Polyhedra", *Mathematical Programming 33*, 1–27. [170]

G. Cornuejols & G.L. Nemhauser (1978), "Tight Bounds for Christofides' TSP Heuristic", *Mathematical Programming 14*, 116–121. [92]

T.M. Cronin (1990), "The Voronoi Diagram for the Euclidean Traveling Salesman Problem is Piecemeal Hyperbolic", Research Report, CECOM Center for Signals Warfare, Warrenton. [48]

H. Crowder & M.W. Padberg (1980), "Solving Large-Scale Symmetric Traveling Salesman Problems to optimality", *Management Science 26*, 495–509. [1, 201]

G.B. Dantzig, D.R. Fulkerson & S.M. Johnson (1954), "Solution of a Large Scale Traveling-Salesman Problem", *Operations Research 2*, 393–410. [1]

G.B. Dantzig (1963), *Linear Programming and Extensions*, Princeton University Press, Princeton. [13]

B. Delaunay (1934), "Sur la sphère vide", *Izvestia Akademia Nauk. SSSr, VII Seria, Otdelenie Matemtischeskii i Estestvennyka Nauk 6*, 793–800. [48]

L. Devroye (1980), "A Note on Finding Convex Hulls via Maximal Vectors", *Information Processing Letters 11*, 53–56. [62]

L. Devroye & G.T. Toussaint (1981), "A Note on Linear Expected Time Algorithms for Finding Convex Hulls", *Computing 26*, 361–366. [58, 62]

M.B. Dillencourt (1987a), "Traveling Salesman Cycles are not Always Subgraphs of Delaunay Triangulations or of Minimum Weight Triangulations", *Information Processing Letters 24*, 339–342. [67]

M.B. Dillencourt (1987b), "A Non-Hamiltonian, Nondegenerate Delaunay Triangulation", *Information Processing Letters 25*, 149–151. [67]

W. Dreissig & W. Uebach (1990), Personal Communication [36]

G. Dueck (1993), "New Optimization Heuristics. The Great-Deluge Algorithm and the Record-to-Record-Travel", *Journal of Computational Physics 104,* , 86–92. [156]

G. Dueck & T. Scheuer (1990), "Threshold Accepting: A General Purpose Optimization Algorithm Superior to Simulated Annealing", *Journal of Computational Physics 90,* , 161–175. [156]

R. Durbin & D. Willshaw (1987), "An Analogue Approach to the Travelling Salesman Problem Using an Elastic Net Method", *Nature 326,* 689–691. [159]

H. Edelsbrunner (1987), *Algorithms in Combinatorial Geometry,* Springer Verlag, Berlin. [42]

J. Edmonds (1965), "Maximum Matching and a Polyhedron with 0,1-Vertices", *Journal of Research of the National Bureau of Standards B 69,* 125–130. [8, 92]

J. Edmonds & E.L. Johnson (1973), "Matching, Euler Tours and the Chinese Postman", *Mathematical Programming 5,* 88–124. [9, 34, 162]

M. Fischetti & P. Toth (1992), "An Additive Bounding Procedure for the Asymmetric Travelling Salesman Problem", *Mathematical Programming 53,* 173–197. [186]

M. Fischetti & P. Toth (1993), "An Efficient Algorithm for the Min-Sum Arborescence Problem on Complete Directed Graphs", *ORSA Journal on Computing 5,* 426–434. [165]

S. Fortune (1987), "A Sweepline Algorithm for Voronoi Diagrams", *Algorithmica 2,* 153–174. [48]

A.M. Frieze (1979), "Worst-case Analysis of Algorithms for Travelling Salesman Problems", *OR Verfahren 32,* 93–112. [97, 164]

B. Fritzke & P. Wilke (1991), "FLEXMAP – A Neural Network for the Traveling Salesman Problem with Linear Time and Space Complexity", Proc. International Joint Conference on Neural Networks, Singapore, 929–934. [159]

B. Fruhwirth (1987), "Untersuchung über genetisch motivierte Heuristiken für das Euklidische Rundreiseproblem", Research Report, TU Graz. [158]

M.R. Garey & D.S. Johnson (1979), *Computers and Intractability: A Guide to the Theory of $\mathcal{NP}$-Completeness,* Freeman, San Francisco. [7]

R.S. Garfinkel (1985), "Motivation and Modeling", in: E.L. Lawler, J.K. Lenstra, A.H.G. Rinnooy Kan & D.B. Shmoys (eds.), *The Traveling Salesman Problem*, John Wiley & Sons, Chichester, 307–360. [31]

B. Gavish & K. Srikanth (1986), "An Optimal Method for Large-Scale Multiple Traveling Salesmen Problems", *Operations Research 34,* 698–717. [34]

M. Gendreau, A. Hertz & G. Laporte (1992), "New Insertion and Postoptimization Procedures for the Traveling Salesman Problem", *Operations Research 40,* 1086–1094. [132]

F. Glover (1989), "Tabu Search", *ORSA Journal on Computing 1,* 190–206 (Part I), *ORSA Journal on Computing 2,* 4–32 (Part II). [159]

F. Glover (1992), "Ejection Chains, Reference Structures, and Alternating Path Methods for the Traveling Salesman Problem", Research Report, University of Colorado, Boulder. [132]

D.E. Goldberg (1989), *Genetic Algorithms in Search, Optimization and Machine Learning,* Addison-Wesley. [158]

B.L. Golden & W.R. Stewart (1985), "Empirical Analysis of Heuristics", in: E.L. Lawler, J.K. Lenstra, A.H.G. Rinnooy Kan & D.B. Shmoys (eds.), *The Traveling Salesman Problem*, John Wiley & Sons, Chichester, 207–249. [73]

M. Golin & R. Sedgewick (1988), "Analysis of a Simple Yet Efficient Convex Hull Algorithm", *Proceedings of the 4th Annual Symposium on Computational Geometry*, 153–163. [62]

M. Gorges-Schleuter (1990), "A Massive Parallel Genetic Algorithm for the TSP", Talk presented at the TSP-Workshop 1990, CRPC, Rice University. [158]

R.L. Graham (1972), "An Efficient Algorithm for Determining the Convex Hull fo a Finite Planar Set", *Information Processing Letters 1,* 132–133. [56]

P.J. Green & R. Sibson (1978), "Computing Dirichlet Tessellations in the Plane", *The Comput. Journal 21,* 168–173. [46]

M. Grötschel (1977), *Polyedrische Charakterisierungen kombinatorischer Optimierungsprobleme*, Hain, Meisenheim am Glan. [201]

M. Grötschel (1980), "On the Symmetric Traveling Salesman Problem: Solution of a 120-city Problem", *Mathematical Programming Studies 12,* 61–77. [1]

M. Grötschel & O. Holland (1991), "Solution of Large-Scale Symmetric Travelling Salesman Problems", *Mathematical Programming 51,* 141–202. [1, 201]

M. Grötschel, M. Jünger & G. Reinelt (1989), "Via Minimization with Pin Preassignments and Layer Preference", *Zeitschrift für Angewandte Mathematik und Mechanik 69,* 393–399. [193]

M. Grötschel, M. Jünger & G. Reinelt (1991), "Optimal Control of Plotting and Drilling Machines: A Case Study", *Zeitschrift für Operations Research – Methods and Models of Operations Research 35,* 61–84. [187, 189, 195, 196, 197]

M. Grötschel, L. Lovász & A. Schrijver (1988), *Geometric Algorithms and Combinatorial Optimization*, Springer, Heidelberg. [163, 203]

M. Grötschel & M.W. Padberg (1985), "Polyhedral theory", in: E.L. Lawler, J.K. Lenstra, A.H.G. Rinnooy Kan & D.B. Shmoys (eds.), *The Traveling Salesman Problem*, John Wiley & Sons, Chichester, 251–305. [163, 202]

B. Hajek (1985), "A Tutorial Survey of Theory and Applications of Simulated Annealing", Proc. 24th IEEE Conf. on Decision and Control, 755–760. [156]

K.H. Helbig-Hansen & J. Krarup (1974), "Improvements of the Held-Karp Algorithm for the Symmetric Traveling Salesman Problem", *Mathematical Programming 7,* 87–96. [176]

M. Held & R.M. Karp (1970), "The Traveling Salesman Problem and Minimum Spanning Trees", *Operations Research 18,* 1138–1162. [176]

M. Held & R.M. Karp (1971), "The Traveling Salesman Problem and Minimum Spanning Trees: Part II", *Mathematical Programming 1,* 6–25. [176]

N. Henriques, F. Safayeni & D. Fuller (1987), "Human Heuristics for the Traveling Salesman Problem", Research Report, University of Waterloo. [159]

A.J. Hoffman & P. Wolfe (1985), "History", in: E.L. Lawler, J.K. Lenstra, A.H.G. Rinnooy Kan & D.B. Shmoys, eds., *The Traveling Salesman Problem*, John Wiley & Sons, Chichester, 1–15. [1]

J.J. Hopfield & D.W. Tank (1985), "Neural Computation of Decisions in Optimization Problems", *Biol. Cybern. 52,* 141–152. [160]

D.J. Houck, J-C. Picard, M. Queyranne & R.R. Vemuganti (1980), "The Travelling Salesman Problem as a Constrained Shortest Path Problem: Theory and Computational Experience", *OPSEARCH 17,* 94–109. [186]

T.C. Hu (1965), "Decomposition in Traveling Salesman Problems", Proc. IFORS Theory of Graphs, A34–A44. [151]

C.A.J. Hurkens (1991), "Nasty TSP Instances for Classical Insertion Heuristics", Research Report, University of Technology, Eindhoven. [84]

D.S. Johnson (1990), "Local Optimization and the Traveling Salesman Problem", in: G. Goos & J. Hartmanis (eds.) *Automata, Languages and Programming*, Lecture Notes in Computer Science 442, Springer, Heidelberg, 446–461. [73, 79, 84, 92, 106, 119, 129, 138, 145, 155]

D.S. Johnson & C.H. Papadimitriou (1985), "Computational Complexity", in: E.L. Lawler, J.K. Lenstra, A.H.G. Rinnooy Kan & D.B. Shmoys (eds.) *The Traveling Salesman Problem*, John Wiley & Sons, Chichester, 37–85. [9, 10, 11]

D.S. Johnson, C.R. Aragon, L.A. McGeoch & C. Schevon (1991), "Optimization by Simulated Annealing: An Experimental Evaluation", *Operations Research 37,* 865–892 (Part I), *Operations Research 39,* 378–406 (Part II). [155]

R. Jonker & T. Volgenant (1988), "An Improved Transformation of the Symmetric Multiple Traveling Salesman Problem", *Operations Research 36,* 163–167. [34]

M. Jünger & W.R. Pulleyblank (1993), "Geometric Duality and Combinatorial Optimization", Jahrbuch Überblicke Mathematik 1993, Vieweg, Braunschweig, 1–24. [168]

M. Jünger, G. Reinelt & G. Rinaldi (1994), "The Traveling Salesman Problem", in: M. Ball, T. Magnanti, C.L. Monma & G. Nemhauser (eds.) *Handbook on Operations Research and Management Sciences: Networks*, North-Holland. [2, 195, 200]

M. Jünger, G. Reinelt & S. Thienel (1993), "Optimal and Provably Good Solutions for the Symmetric Traveling Salesman Problem", to appear in *Zeitschrift für Operations Research*. [184, 200, 209, 210]

M. Jünger, G. Reinelt & S. Thienel (1994), "Practical Problem Solving with Cutting Plane Algorithms in Combinatorial Optimization", to appear in:*DIMACS Series in Discrete Mathematics and Theoretical Computer Science*. [200]

M. Jünger, G. Reinelt & D. Zepf (1991), "Computing Correct Delaunay Triangulations", *Computing 47,* 43–49. [50]

V. Kaibel (1993), *Numerisch stabile Berechnung von Voronoi-Diagrammen*, Diplomarbeit, Universität zu Köln. [52]

N. Karmarkar (1984), "A New Polynomial-time Algorithm for Linear Programming", *Combinatorica 4,* 373–395. [13]

R. Karp (1977), "Probabilistic Analysis of Partitioning Algorithms for the Traveling-Salesman in the Plane", *Mathematics of Operations Research 2,* 209–224. [151]

R.M. Karp & J.M. Steele (1985), "Probabilistic Analysis of Heuristics", in: E.L. Lawler, J.K. Lenstra, A.H.G. Rinnooy Kan & D.B. Shmoys (eds.) *The Traveling Salesman Problem*, John Wiley & Sons, Chichester, 181–205. [151]

C. Kemke (1988), "Der Neuere Konnektionismus; Ein Überblick", *Informatik Spektrum 11,* 143–162. [160]

L.G. Khachiyan (1979), "A Polynomial Algorithm in Linear Programming", *Soviet Mathematics Doklady 20,* 191–194). [13]

S. Kirkpatrick (1984), "Optimization by Simulated Annealing: Quantitative Studies", *Journal of Statistical Physics 34,* 975–986. [155]

S. Kirkpatrick, C.D. Gelatt Jr. & M.P. Vecchi (1983), "Optimization by Simulated Annealing", *Science 222,* 671–680. [154]

D.G. Kirkpatrick & R. Seidel (1986), "The Ultimate Planar Convex Hull Algorithm", *SIAM Journal on Computing 15,* 287–299. [58, 62]

K.C. Kiwiel (1989), "A Survey of Bundle Methods for Nondifferentiable Optimization", in: M. Iri & K. Tanabe (eds.) *Mathematical Programming. Recent Developments and Applications*, Kluwer Academic Publishers, Dordrecht, 263–282. [186]

J. Knox & F. Glover (1989), "Comparative Testing of Traveling Salesman Heuristics Derived from Tabu Search, Genetic Algorithms and Simulated Annealing", Research Report, University of Colorado. [159]

D.E. Knuth (1973), "*The Art of Computer Programming, Volume 3: Sorting and Searching*", Addison-Wesley, Reading. [14]

G. Kolata (1991), "Math Problem, Long Baffling, Slow Yields", The New York Times, March 12. [1]

B. Korte (1989), "Applications of Combinatorial Optimization", in: M. Iri & K. Tanabe (eds.) *Mathematical Programming. Recent Developments and Applications*, Kluwer Academic Publishers, Dordrecht, 203–225. [194, 195]

J.B. Kruskal (1956), "On the Shortest Spanning Subtree of a Graph and the Traveling Salesman Problem", *Proceedings of the American Mathematical Society 7,* 48–50. [29]

H.T. Kung, F. Luccio & F.P. Preparata (1975), "On Finding the Maxima of a Set of Vectors", *Journal of the ACM 4,* 469–476. [55, 59]

E.L. Lawler, J.K. Lenstra, A.H.G. Rinnooy Kan & D.B. Shmoys (eds.) (1985), *The Traveling Salesman Problem,* John Wiley & Sons, Chichester. [2]

D.T. Lee & F.P. Preparata (1984), "Computational Geometry — A Survey", *IEEE Transactions on Computers C-33,* 1072–1101. [62]

J.K. Lenstra & A.H.G. Rinnooy Kan (1974), "Some Simple Applications of the Travelling Salesman Problem", Research Report, Stichting Mathematisch Centrum, Amsterdam. [37, 38, 39]

S. Lin & B.W. Kernighan (1973), "An Effective Heuristic Algorithm for the Traveling-Salesman Problem", *Operations Research 21,* , 498–516. [123]

J.D. Litke (1984), "An Improved Solution to the Traveling Salesman Problem with Thousands of Nodes", *Communications of the ACM 27,* 1227–1236. [145]

I.J. Lustig, R.E. Marsten & D.F. Shanno (1994), "Interior Point Methods for Linear Programming: Computational State of the Art", *ORSA Journal on Computing 6,* 1–14. [13]

N. Maculan & J.J.C. Salles (1989), "A Lower Bound for the Shortest Hamiltonean Path Problem in Directed Graphs", Research Report, COPPE, Rio de Janeiro. [186]

K.-T. Mak & A.J. Morton (1993), "A Modified Lin-Kernighan Traveling Salesman Heuristic", *Operations Research Letters 13,* 127–132. [124]

M. Malek, M. Guruswamy, H. Owens & M. Pandya (1989), "Serial and Parallel Search Techniques for the Traveling Salesman Problem", *Annals of OR: Linkages with Artificial Intelligence.* [159]

M. Malek, M. Heap, R. Kapur & A. Mourad (1989), "A Fault Tolerant Implementation of the Traveling Salesman Problem", Research Report, University of Texas at Austin. [159]

K. Malik & M.L. Fisher (1990), "A Dual Ascent Algorithm for the 1-Tree Relaxation of the Symmetric Traveling Salesman Problem", *Operations Research Letters 9,* 1–7. [179]

F. Margot (1992), "Quick Updates for p-OPT TSP heuristics", *Operations Research Letters 11.* [123]

O. Martin, S.W. Otto & E.W. Felten (1992), "Large-step Markov Chains for the TSP incorporating local search heuristics", *Operations Research Letters 11,* 219–224. [156]

N. Metropolis, A. Rosenbluth, M. Rosenbluth, A. Teller & E. Teller (1953), "Equation of State Calculation by Fast Computing Machines", *Journal Chem. Physics 21,* 1087–1092. [153]

D.L. Miller & J.F. Pekny (1991), "Exact Solution of Large Asymmetric Traveling Salesman Problems", *Science 251,* 754–761. [201]

D.L. Miller, J.F. Pekny & G.L. Thompson (1991), "An Exact Two-Matching Based Branch and Bound Algorithm for the Symmetric Traveling Salesman Problem", Research Report, Carnegie Mellon University, Pittsburgh. [201]

H. Mühlenbein, M. Gorges-Schleuter & O. Krämer (1988), "Evolution Algorithms in Combinatorial Optimization", *Parallel Computing 7,* 65–85. [158]

D. Naddef & G. Rinaldi (1991), "The Symmetric Traveling Salesman Polytope and its Graphical Relaxation: Composition of Valid Inequalities", *Mathematical Programming 51, 359-400.* [202]

D. Naddef & G. Rinaldi (1993), "The Graphical Relaxation: A New Framework for the Symmetric Traveling Salesman Polytope", *Mathematical Programming* 58, 53–88. [32, 202]

G.L. Nemhauser & L.A. Wolsey (1988), *Integer and Combinatorial Optimization,* John Wiley & Sons, Chichester. [12]

T. Ohya, M. Iri & K. Murota (1984), "Improvements of the Incremental Method for the Voronoi Diagram with Computational Comparison of Various Algorithms", *Journal of the Operations Research Society of Japan 27,* 306–337. [47, 48, 52]

I. Or (1976), *Traveling Salesman-Type Combinatorial Problems and Their Relation to the Logistics of Regional Blood Banking,* Doctoral Thesis, Northwestern University, Evanston. [122]

T. Ottmann & P. Widmayer (1990), *Algorithmen und Datenstrukturen,* BI-Wissenschaftsverlag, Mannheim. [44, 115]

M.W. Padberg & M. Grötschel (1985), "Polyhedral Computations", in: E.L. Lawler, J.K. Lenstra, A.H.G. Rinnooy Kan & D.B. Shmoys (eds.), *The Traveling Salesman Problem,* John Wiley & Sons, Chichester, 307–360. [163]

M.W. Padberg & G. Rinaldi (1987), "Optimization of a 532 City Symmetric Traveling Salesman Problem by Branch and Cut", *Operations Research Letters 6,* 1–7. [1]

M.W. Padberg & G. Rinaldi (1991), "A Branch & Cut Algorithm for the Resolution of Large-scale Symmetric Traveling Salesman Problems", *SIAM Review 33,* 60–100. [1, 201]

J.F. Pekny & D.L. Miller (1994), "A Staged Primal-Dual Algorithm for Finding a Minimum Cost Perfect 2-Matching in an Undirected Graph", *ORSA Journal on Computing 6,* 68–81. [162]

J. Perttunen (1991), "On the Significance of the Initial Solution in Edge Exchange Heuristics", Research Report, University of Vaasa. [132]

R.D. Plante, T.J. Lowe & R. Chandrasekaran (1987), "The Product Matrix Traveling Salesman Problem: An Application and Solution Heuristics", *Operations Research 35,* 772–783. [36]

B.T. Polyak (1978), *Subgradient Methods: A Survey of Soviet Research,* in: C. Lemarèchal & R. Mifflin (eds.), *Nonsmooth Optimization,* Pergamon Press, Oxford, 5–29. [174]

J.-Y. Potvin (1993), "The Traveling Salesman Problem: A Neural Network Perspective", *ORSA Journal On Computing 5,* 328–348. [159]

J.-Y. Potvin & J.-M. Rousseau (1990), "Enhancements to the Clarke and Wright Algorithm for the Traveling Salesman Problem", Research Report, University of Montreal. [94]

F.P. Preparata & S.J. Hong (1977), "Convex Hulls of Finite Sets of Points in Two and Three Dimensions", *Communications of the ACM 20,* 87–93. [55]

R.C. Prim (1957), "Shortest Connection Networks and Some Generalizations", *The Bell System Technical Journal 36,* 1389–1401. [28, 91]

R. Ramesh, Y.-S. Yoon & M.H. Karwan (1992), "An Optimal Algorithm for the Orienteering Tour Problem", *ORSA Journal on Computing 4,* 155–165. [35]

H.D. Ratliff & A.S. Rosenthal (1981), "Order-Picking in a Rectangular Warehouse: A Solvable Case for the Travelling Salesman Problem", *Operations Research 31,* 507–521. [36]

I. Rechenberg (1973), *Evolutionsstrategie: Optimierung technischer Systeme nach Prinzipien der biologischen Evolution,* Frommann-Holzboog, Stuttgart. [157]

G. Reinelt (1991a), "TSPLIB – A Traveling Salesman Problem Library", *ORSA Journal on Computing 3,* 376–384. [41, 211, 212]

G. Reinelt (1991b), *TSPX – A Software Package for Practical Traveling Salesman Problem Solving,* Universität Augsburg, Augsburg, 1991. [14]

D.J. Rosenkrantz, R.E. Stearns & P.M. Lewis (1977), "An Analysis of Several Heuristics for the Traveling Salesman Problem", *SIAM Journal on Computing 6,* 563–581. [74, 84, 123]

P. Ruján (1988), "Searching for Optimal Configurations by Simulated Tunneling", *Zeitschrift für Physik B – Condensed Matter 73,* 391–416. [156]

P. Ruján, C. Evertsz & J.W. Lyklema (1988), "A Laplacian Walk for the Travelling Salesman", *Europhysics Letters 7,* 191–195. [48]

D.E. Rumelhart, G.E. Hinton & J.L. McClelland (1986), *The PDP Research Group: Parallel Distributed Processing: Explorations in the Microstructure of Cognition,* MIT Press. [160]

S. Sahni & T. Gonzales (1976), "P-complete Approximation Problems", *Journal of the ACM 23,* 555-565. [11]

H. Schramm (1989), *Eine Kombination von Bundle- und Trust-Region-Verfahren zur Lösung nichtdifferenzierbarer Optimierungsprobleme,* Bayreuther Mathematische Schriften, Heft 30. [186]

A. Schrijver (1986), *Theory of Linear and Integer Programming,* John Wiley & Sons, Chichester. [12, 174]

M.I. Shamos & D. Hoey (1975), "Closest Point Problems", Proc. 16th IEEE Annual Symposium on Foundations of Computer Science, 151–162. [45]

D.B. Shmoys & D.P. Williamson (1990), "Analyzing the Held-Karp TSP Bound: A Monotonicity Property with Application", *Information Processing Letters 35,* 281–285. [184]

T.H.C. Smith & G.L. Thompson (1977), "A LIFO Implicit Enumeration Search Algorithm for the Symmetric Traveling Salesman Problem Using Held & Karp's 1-Tree Relaxation", *Annals of Discrete Mathematics 1,* 479–493. [176]

T.H.C. Smith, T.W.S. Meyer & G.L. Thompson (1990), "Lower Bounds for the Symmetric Traveling Salesman Problem from Lagrangean Relaxations", *Discrete Applied Mathematics 26,* 209–217. [182, 184]

K. Sugihara (1988), "A Simple Method for Avoiding Numerical Errors and Degeneracy in Voronoi Diagram Construction", Research Memorandum, University of Tokyo. [50]

K. Sugihara & M. Iri (1988), "Geometric Algorithms in Finite-Precision Arithmetic", Research Memorandum, University of Tokyo. [50]

R.E. Tarjan (1983), *Data Structures and Network Algorithms,* Society for Industrial and Applied Mathematics. [14, 18, 23]

S. Thienel (1991), *Schnelle Berechnung von konvexen Hüllen im dreidimensionalen euklidischen Raum,* Diplomarbeit, Universität Augsburg. [62]

G.T. Toussaint (1985), "A Historical Note on Convex Hull Finding Algorithms", *Pattern Recognition Letters 3,* 21–28. [55]

N.L.J. Ulder, E. Pesch, P.J.M. van Laarhoven, H.-J. Bandelt & E.H.L. Aarts (1990), "Improving TSP Exchange Heuristics by Population Genetics", Research Report, Erasmus Universiteit Rotterdam. [158]

R. van Dal (1992), *Special Cases of the Traveling Salesman Problem,* Wolters-Noordhoff, Groningen. [11, 37]

R. van der Veen (1992), *Solvable Cases of the Traveling Salesman Problem with Various Objective Functions,* Doctoral Thesis, Rijksuniversiteit Groningen, Groningen. [11]

P.J.M. van Laarhoven (1988), *Theoretical and Computational Aspects of Simulated Annealing,* Doctoral Thesis, Erasmus Universiteit Rotterdam. [155]

T. Volgenant & R. Jonker (1982), "A Branch and Bound Algorithm for the Symmetric Traveling Salesman Problem Based on the 1-Tree Relaxation", *European Journal of Operations Research 9*, 83–89. [176]

G. Voronoi (1908), "Nouvelles Applications des Paramètres Continus à la Théorie des Formes Quadratiques. Deuxième Memoire: Recherche sur les Parallélloèdres Primitifs", *Journal für Reine und Angewandte Mathematik 3*, 198–287. [42]

R.H. Warren (1993), "Special Cases of the Traveling Salesman Problem", to appear in *Applied Mathematics and Computation.* [11]

M. Wottawa (1991), *Parallelisierung von Heuristiken für das Traveling Salesman Problem und ihre Implementierung auf einem Transputer-Cluster,* Diplomarbeit, Universität zu Köln. [152]

Index

Springer-Verlag and the Environment

We at Springer-Verlag firmly believe that an international science publisher has a special obligation to the environment, and our corporate policies consistently reflect this conviction.

We also expect our business partners – paper mills, printers, packaging manufacturers, etc. – to commit themselves to using environmentally friendly materials and production processes.

The paper in this book is made from low- or no-chlorine pulp and is acid free, in conformance with international standards for paper permanency.

Lecture Notes in Computer Science

For information about Vols. 1–762
please contact your bookseller or Springer-Verlag

Vol. 800: J.-O. Eklundh (Ed.), Computer Vision – ECCV '94. Proceedings 1994, Vol. I. XVIII, 603 pages. 1994.

Vol. 801: J.-O. Eklundh (Ed.), Computer Vision – ECCV '94. Proceedings 1994, Vol. II. XV, 485 pages. 1994.

Vol. 802: S. Brookes, M. Main, A. Melton, M. Mislove, D. Schmidt (Eds.), Mathematical Foundations of Programming Semantics. Proceedings, 1993. IX, 647 pages. 1994.

Vol. 803: J. W. de Bakker, W.-P. de Roever, G. Rozenberg (Eds.), A Decade of Concurrency. Proceedings, 1993. VII, 683 pages. 1994.

Vol. 804: D. Hernández, Qualitative Representation of Spatial Knowledge. IX, 202 pages. 1994. (Subseries LNAI).

Vol. 805: M. Cosnard, A. Ferreira, J. Peters (Eds.), Parallel and Distributed Computing. Proceedings, 1994. X, 280 pages. 1994.

Vol. 806: H. Barendregt, T. Nipkow (Eds.), Types for Proofs and Programs. VIII, 383 pages. 1994.

Vol. 807: M. Crochemore, D. Gusfield (Eds.), Combinatorial Pattern Matching. Proceedings, 1994. VIII, 326 pages. 1994.

Vol. 808: M. Masuch, L. Pólos (Eds.), Knowledge Representation and Reasoning Under Uncertainty. VII, 237 pages. 1994. (Subseries LNAI).

Vol. 809: R. Anderson (Ed.), Fast Software Encryption. Proceedings, 1993. IX, 223 pages. 1994.

Vol. 810: G. Lakemeyer, B. Nebel (Eds.), Foundations of Knowledge Representation and Reasoning. VIII, 355 pages. 1994. (Subseries LNAI).

Vol. 811: G. Wijers, S. Brinkkemper, T. Wasserman (Eds.), Advanced Information Systems Engineering. Proceedings, 1994. XI, 420 pages. 1994.

Vol. 812: J. Karhumäki, H. Maurer, G. Rozenberg (Eds.), Results and Trends in Theoretical Computer Science. Proceedings, 1994. X, 445 pages. 1994.

Vol. 813: A. Nerode, Yu. N. Matiyasevich (Eds.), Logical Foundations of Computer Science. Proceedings, 1994. IX, 392 pages. 1994.

Vol. 814: A. Bundy (Ed.), Automated Deduction—CADE-12. Proceedings, 1994. XVI, 848 pages. 1994. (Subseries LNAI).

Vol. 815: R. Valette (Ed.), Application and Theory of Petri Nets 1994. Proceedings. IX, 587 pages. 1994.

Vol. 816: J. Heering, K. Meinke, B. Möller, T. Nipkow (Eds.), Higher-Order Algebra, Logic, and Term Rewriting. Proceedings, 1993. VII, 344 pages. 1994.

Vol. 817: C. Halatsis, D. Maritsas, G. Philokyprou, S. Theodoridis (Eds.), PARLE '94. Parallel Architectures and Languages Europe. Proceedings, 1994. XV, 837 pages. 1994.

Vol. 818: D. L. Dill (Ed.), Computer Aided Verification. Proceedings, 1994. IX, 480 pages. 1994.

Vol. 819: W. Litwin, T. Risch (Eds.), Applications of Databases. Proceedings, 1994. XII, 471 pages. 1994.

Vol. 820: S. Abiteboul, E. Shamir (Eds.), Automata, Languages and Programming. Proceedings, 1994. XIII, 644 pages. 1994.

Vol. 821: M. Tokoro, R. Pareschi (Eds.), Object-Oriented Programming. Proceedings, 1994. XI, 535 pages. 1994.

Vol. 822: F. Pfenning (Ed.), Logic Programming and Automated Reasoning. Proceedings, 1994. X, 345 pages. 1994. (Subseries LNAI).

Vol. 823: R. A. Elmasri, V. Kouramajian, B. Thalheim (Eds.), Entity-Relationship Approach — ER '93. Proceedings, 1993. X, 531 pages. 1994.

Vol. 824: E. M. Schmidt, S. Skyum (Eds.), Algorithm Theory – SWAT '94. Proceedings. IX, 383 pages. 1994.

Vol. 825: J. L. Mundy, A. Zisserman, D. Forsyth (Eds.), Applications of Invariance in Computer Vision. Proceedings, 1993. IX, 510 pages. 1994.

Vol. 826: D. S. Bowers (Ed.), Directions in Databases. Proceedings, 1994. X, 234 pages. 1994.

Vol. 827: D. M. Gabbay, H. J. Ohlbach (Eds.), Temporal Logic. Proceedings, 1994. XI, 546 pages. 1994. (Subseries LNAI).

Vol. 828: L. C. Paulson, Isabelle. XVII, 321 pages. 1994.

Vol. 829: A. Chmora, S. B. Wicker (Eds.), Error Control, Cryptology, and Speech Compression. Proceedings, 1993. VIII, 121 pages. 1994.

Vol. 830: C. Castelfranchi, E. Werner (Eds.), Artificial Social Systems. Proceedings, 1992. XVIII, 337 pages. 1994. (Subseries LNAI).

Vol. 831: V. Bouchitté, M. Morvan (Eds.), Orders, Algorithms, and Applications. Proceedings, 1994. IX, 204 pages. 1994.

Vol. 832: E. Börger, Y. Gurevich, K. Meinke (Eds.), Computer Science Logic. Proceedings, 1993. VIII, 336 pages. 1994.

Vol. 833: D. Driankov, P. W. Eklund, A. Ralescu (Eds.), Fuzzy Logic and Fuzzy Control. Proceedings, 1991. XII, 157 pages. 1994. (Subseries LNAI).

Vol. 834: D.-Z. Du, X.-S. Zhang (Eds.), Algorithms and Computation. Proceedings, 1994. XIII, 687 pages. 1994.

Vol. 835: W. M. Tepfenhart, J. P. Dick, J. F. Sowa (Eds.), Conceptual Structures: Current Practices. Proceedings, 1994. VIII, 331 pages. 1994. (Subseries LNAI).

Vol. 836: B. Jonsson, J. Parrow (Eds.), CONCUR '94: Concurrency Theory. Proceedings, 1994. IX, 529 pages. 1994.

Vol. 837: S. Wess, K.-D. Althoff, M. M. Richter (Eds.), Topics in Case-Based Reasoning. Proceedings, 1993. IX, 471 pages. 1994. (Subseries LNAI).

Vol. 838: C. MacNish, D. Pearce, L. Moniz Pereira (Eds.), Logics in AI. Proceedings, 1994. IX, 413 pages. 1994. (Subseries LNAI).

Vol. 839: Y. G. Desmedt (Ed.), Advances in Cryptology - CRYPTO '94. Proceedings, 1994. XII, 439 pages. 1994.

Vol. 840: G. Reinelt, The Traveling Salesman. VIII, 223 pages. 1994.

Vol. 841: I. Prívara, B. Rovan, P. Ružička (Eds.), Mathematical Foundations of computer Science 1994. Proceedings, 1994. X, 628 pages. 1994.

Vol. 842: T. Kloks, Treewidth. IX, 209 pages. 1994.

Vol. 843: A. Szepietowski, Turing Machines with Sublogarithmic Space. VIII, 115 pages. 1994.